Recent Advances in Botanical Science

(*Volume 1*)

Contemporary Research on Bryophytes

Edited by
Afroz Alam
*Department of Bioscience and Biotechnology, Banasthali Vidyapith,
Rajasthan- 304022, India*

Recent Advances in Botanical Science

Volume # 1

Contemporary Research on Bryophytes

Editor: Afroz Alam

ISBN (Online): 978-981-14-3378-8

ISBN (Print): 978-981-14-3376-4

© 2020, Bentham Books imprint.

Published by Bentham Science Publishers Pte. Ltd. Singapore. All Rights Reserved.

need for a court order if at any point you breach any terms of this License Agreement. In no event will any delay or failure by Bentham Science Publishers in enforcing your compliance with this License Agreement constitute a waiver of any of its rights.

3. You acknowledge that you have read this License Agreement, and agree to be bound by its terms and conditions. To the extent that any other terms and conditions presented on any website of Bentham Science Publishers conflict with, or are inconsistent with, the terms and conditions set out in this License Agreement, you acknowledge that the terms and conditions set out in this License Agreement shall prevail.

Bentham Science Publishers Pte. Ltd.
80 Robinson Road # 02-00
Singapore 068898
Singapore
Email: subscriptions@benthamscience.net

BENTHAM SCIENCE

CONTENTS

FOREWORD

The book titled: 'Contemporary Research on Bryophytes' is edited by Dr Afroz Alam, who received his initial training in bryological research at the Department of Botany, University of Lucknow, Lucknow - an internationally acclaimed and foremost centre of research in the field of plant sciences in India for nearly eight decades. This book is an extremely remarkable effort for compilation of the diverse aspects of bryology to 12 articles contributed by Indian as well as overseas bryologists specially from Brazil and Pakistan, covering the current status of bryological research in these countries. The areas covered are related to contributions to the conservation of endangered bryophytes through tissue culture, the present status of Antarctic bryology, application of molecular approaches and bioinformatics to study bryophytes besides, morphology, taxonomy and bryodiversity. It provides information on both the modern aspects of research in bryology and also the classical aspect which is commendable on the part of the contributors. I hope that the book will benefit the researchers and serve as a reference resource in the specialisation of plant sciences, an area of research not so popularly known in this part of the world.

S C Srivastava
Former Professor & Head
Department of Botany
University of Lucknow
Lucknow (U.P.)
India

PREFACE

Bryophytes are the foremost green plants to effectively inhabit land 470–551 million years ago (mya) from algal intimates. Conventionally, the word "bryophytes" has been used to explain a paraphyletic assembly of three phyla, Bryophyta (mosses), Marchantiophyta (liverworts), and Anthocerotophyta (hornworts) and their relations have long been a cause of arguments among researchers. Being the first land plants and one of the pioneers along with lichens, they have always fascinated botanists, and consequently the research on this group of miniature plants is global. This book is conceptualized to provide recent trends in bryology to the scientific community worldwide. The study of bryophytes is a basic in land plant evolution, and the most recent progress in molecular and computational biology has permitted for a precise depiction of land colonization and the succeeding evolution. Ecologically, bryophytes are very important and serve as water reservoirs along with their active participation in extremely important biogeochemical cycles, particularly the carbon cycle. Furthermore, there has been an increasing concern for the conservation of bryophytes.

The aim of this book is to present an all-inclusive text concerning current research set-up of bryology to bryologists.

This book is a collection of chapters written by renowned bryologists from Brazil, Pakistan and India who reported findings on diverse aspects of bryology, which include, diversity, conservation through tissue culture, pollution monitoring, bioinformatics, molecular bryology, *etc.*

The text has been written in a simple graspable language and is profusely illustrated with self-explanatory diagrams.

Suggestions and healthy criticisms will be of much help and value to me and shall be incorporated in the future attempts.

Afroz Alam
Department of Bioscience and Biotechnology,
Banasthali Vidyapith,
Rajasthan- 304022,
India

INTRODUCTION

The aim of the book is to provide information regarding recent trends in bryology in India and abroad. Bryophytes are the first land plants and are quite fascinating in their overall diversity. Due to the lack of recent relevant research on their mesmerizing diversity and evolutionary history, there is a need to know the advancement in this branch of plant sciences. All through its history, bryological study has contributed considerably to the field of plant sciences, for instance, the discovery of sex chromosomes in plants. The study of bryophytes is fundamental in land plant evolution, and the latest progress in molecular phylogenetics and genomics has permitted for an exact depiction of land immigration and succeeding progression to emerge. Ecologically, bryophytes are important for the participation in biogeochemical cycles, in particular carbon cycle. Furthermore, there has been an escalating apprehension for the conservation biology of bryophytes.

This book intends to put forward an all-inclusive text regarding the current research scenario of bryology to those interested in finding out the fundamentals of the bryophytes.

This book is a compilation of chapters on reported findings related to various aspects of bryology like, conservation, diversity, tissue culture, bio-monitoring, computational bryology, molecular bryology, *etc.*

The purpose of the book is to provide botanists and bryologists with basic updated and recent information that will be valuable in their quest to investigate, develop, and apply to their research work. An attempt has been made to explain the current advances made across the diverse aspects of bryophytes at a level that would be helpful to beginners in the field.

List of Contributors

Abhishek Tripathi	PCS, Government of Uttarakhand, Uttarakhand, India
Afroz Alam	Department of Bioscience and Biotechnology, Banasthali Vidyapith, Rajasthan- 304022, India
Ashish K. Asthana	Bryology Laboratory, CSIR-National Botanical Research Institute, Lucknow-226001, India
Asheesh Shanker	Department of Bioinformatics, Central University of South Bihar, Gaya-824236, India
A.E. Dulip Daniels	Bryology Laboratory, Department of Botany & Research Centre, Scott Christian College, (Autonomous), Nagercoil - 629 003, Tamil Nadu, India
V.K. Chandini	PG & Research Department of Botany, The Zamorin's Guruvayurappan College, Kozhikode-14, Kerala, India
Dayanidhi Gupta	Bryology Laboratory, CSIR-National Botanical Research Institute, Lucknow-226001, India
Dharmendra G. Shah	Ecotoxicology & Lower Plants Lab., Department of Botany, The Maharaja Sayajirao University of Baroda, Vadodara 390 002, Gujarat, India
Jair Putzke	University of UNIPAMPA, Av. Antonio Trilha, 1847 - São Clemente, São Gabriel – RS, CEP: 97300-000, Brazil
Krishna Kumar Rawat	CSIR-National Botanical Research Institute, Lucknow, India
Manju C. Nair	PG & Research Department of Botany, The Zamorin's Guruvayurappan College, Kozhikode-14, Kerala, India
Mazhar-ul- Islam	Cryptogamic Lab. Department of Botany, Hazara University, Mansehra, Pakistan
Praveen Kumar Verma	Forest Botany Division, Forest Research Institute, Dehradun, Uttarakhand, India
K.P. Rajesh	PG & Research Department of Botany, The Zamorin's Guruvayurappan College, Kozhikode-14, Kerala, India
Rajendra Ananda Lavate	Raje Ramrao Mahavidhyalaya, Jath-416 404; Dist. Sangli (M.S.), India
Rakesh V. Gujar	Ecotoxicology & Lower Plants Lab., Department of Botany, The Maharaja Sayajirao University of Baroda, Vadodara 390 002, Gujarat, India
Saumya Pandey	Department of Bioscience and Biotechnology, Banasthali Vidyapith, Rajasthan- 304022, India
Shiv Charan Sharma	Department of Bioscience and Biotechnology, Banasthali Vidyapith, Rajasthan- 304022, India
Sonu Kumar	Department of Bioinformatics, Central University of South Bihar, Gaya-824236, India
Sonu Yadav	Department of Botany, University of Lucknow, Lucknow, India
Vishwa J. Singh	Bryology Laboratory, CSIR-National Botanical Research Institute, Lucknow-226001, India
Vinay Sahu	Bryology Laboratory, CSIR-National Botanical Research Institute, Lucknow-226001, India

In vitro Growth Pattern of Moss *Drummondia stricta* (Mitt.) Müll. Hal. (Orthotrichaceae) in Different Hormonal Concentrations

Vishwa J. Singh, Dayanidhi Gupta, Vinay Sahu and **Ashish K. Asthana**[*]

Bryology Laboratory, CSIR-National Botanical Research Institute, Lucknow-226001, India

Abstract: The present work was undertaken to study the growth pattern of *Drummondia stricta* (Mitt.) Müll. Hal. in six different culture media with a combination of auxin and cytokinins, using spores as explants in aseptic cultures. In *Drummondia,* there is a tendency of multicellular spore formation and spore germination is precocious and endosporic. It has been observed that plant growth was the most satisfactory in the ½ Knop's + 0.1 mg/L IAA + 0.1 mg/L Kinetin followed by ½ Knop's, 0.1 mg/L IAA + 0.1 mg/L BAP. However, no growth was observed in case of ½ KNOP's and Hoagland media.

Keywords: Auxin, Cytokinins, Endosporic, Multicellular spores.

INTRODUCTION

Genus *Drummondia* belongs to the family Orthotrichaceae characterized by creeping plants with numerous erect branches. Capsule erect, ovate, exserted with reduced peristome of 16 teeth and spores multicellular, rectangular to polygonal (Plate **1**). Vitt [1] described six species and one variety of *Drummondia* from World. In India 4 species of *Drummondia* were reported by Chopra [2]. Earlier some research work related to the germination of bryophyte spores, their stages of development, physical factors, nutritional requirements and effects of auxin and antiauxin has been carried out by various workers [3 - 10]. In bryophytes there are two types of spore germination (1) Endosporic and (2) Exosporic. In endosporic germination, the protonema develop inside the stretched spore wall and the spore protoplast divides within the spore wall forming a multicellular structure while in exosporic germination, swollen protoplast ruptures the spore wall and protonema develops outside. Endosporic germination occurs in some epiphytic and saxicolous species [11]. The aims of the present study were (1) to describe the

---

[*] **Corresponding author Ashish K. Asthana:** Bryology Laboratory, CSIR-National Botanical Research Institute, Lucknow-226001, India; Tel: +91-9415105620; E-mail: drakasthana@rediffmail.com

Afroz Alam (Ed.)

spore germination morphology, detailing the phases of *in vitro* protonema development of *D. stricta*, and (2) To evaluate the growth of plant in different culture media.

Fig (1). A-K: *Drummondia stricta*, A. Plant habit, B-C. Leaves, D. Leaf apical cells, E. Leaf median cells, F. Leaf basal cells, G. Capsule, H. SEM of Spores, I-J. SEM of single spore, K. A portion of spore (enlarged view).

MATERIALS AND METHODS

The plants of *D. stricta* were obtained from Govind Wildlife Sanctuary, Uttarkashi, Uttarakhand. The specimens have been deposited in the Bryophyte Herbarium of CSIR-National Botanical Research Institute, Lucknow (LWG), India. Specimens examined: India, western Himalaya, Uttarakhand, Uttarkashi, Obra Jeri (alt. ca 2768 m), Epiphytic, 16.10.2016, leg., Dayanidhi Gupta, 306366A (LWG).

The capsules were surface sterilized with 0.5% sodium hypochlorite solution for 2 minutes, and washed repeatedly with sterile double distilled water. The capsules were ruptured and spores were inoculated in different media. The pH of the media was maintained at 5.8 pH before autoclaving. The media was autoclaved at 15 psi for 15 minutes.

The experiment was carried out in a laboratory under controlled temperature (20-23°C) and provided with illumination of 2400-2500 lux as well as alternate light and dark period of 16 h and 8 h, respectively with the help of a combination of fluorescent tubes.

In order to observe the influence of different mineral nutrients and hormone concentration on the development of this species, the following combinations were used. All the hormone concentrations were prepared in ½ Knop's macronutrient.

(A) 0.1 mg/L IBA + 0.2 mg/L BAP + ½ Knop's

(B) 0.2 mg/L IBA + 0.2 mg/L BAP + ½ Knop's

(C) 0.1 mg/L 2,4 D + 0.1 mg/L Kinetin + ½ Knop's

(D) 0.1 mg/L IAA + 0.1 mg/L BAP + ½ Knop's

(E) Hoagland

(F) ½ Knop's

RESULTS AND DISCUSSION

Germination Stages and Development

The spores of *D. stricta* are multicellular with bright green colour with an average size of spores ranging around 80-120 μm long and 60-80 μm wide. In the first week of the analysis, there was development of multicellular protonema enclosed within the spore wall that expanded without rupturing, like a typical endosporic germination. Germination was observed in A, B, C and D media while in the case of Hoagland and ½ Knop's media only spores were swollen but no further growth occurred. Observation on the development of protonema and sporeling in the second week revealed that in A, B, C and D media massive protonema developed from spores except for Hoagland and ½ Knop's media. During the fourth week spores and protonema became brown in colour in A, B, C, and D media while in the Hoagland and ½ Knop's media dead spores were observed (Table **1**, Plate **2**). The differentiation processes in mosses are clearly dependent on the amount of auxin and cytokinin [12]. In the present experiment cytokinin and auxin accumulation at higher concentrations in the spores inhibited the protonemal growth in A and B media. There are various reports about the effect of auxin and cytokinin on bryophytes, 2, 4 D and 6 Benzyl Amino Purine are usually effective for callus inducement [9, 13, 14]. During the fourth week, leafy buds were observed along the protonema in D media. During the sixth week, leafy bud developed from protonema in C media. Brownish rhizoids also developed from the base of the plants in C and D media. The number of young buds developed more in D media as compared to C media. The plant growth was found best in the case of D media (Table **1**, Plate **3**). Cvetic *et al* [15] reported the Auxin IAA favoured budding and the cytokinin 6 Benzyl Amino purine restrained gametophores growth. Both hormones could accelerate protonemal aging in long culture media.

Table 1. Observations on comparative protonemal morphogenesis and growth pattern of *D. stricta* in different culture media.

Days	Culture Media					
	½ Knop's 0.1 mg/L IBA+ 0.2 mg/L BAP (A)	½ Knop's 0.2 mg/L IBA+0.2 mg/L BAP (B)	½ Knop's 0.1 mg/L 2,4D+0.1 mg/L Kinetin (C)	½ Knop's 0.1 mg/L IAA +0.1 mg/L BAP (D)	Hoagland	½ Knop's
(6th day)	73.93% spores germinated, Spores swollen and growth of protonema started	97.68% spores germinated, Spore swollen and growth of protonema started	98.10% spores germinated, Spore swollen and growth of protonema started	92.67% spores germinated, Spore swollen and growth of protonema started	Spores swollen, and no growth of protonema	Spores swollen, and no growth of protonema
(15th day)	Protonema developed in stretched exospores	Protonema developed in stretched exospores	Protonema developed in stretched exospores	Protonema developed in stretched exospores	no growth of protonema	no growth of protonema
(30th day)	Protonema become brown in color	Protonema become brown in color	Protonema become brown in color No growth of caulonema	Protonema become brown in color No growth of caulonema, leafy bud developed from protonema	Death of Spores	Death of Spores
(45th day)	No further growth	No further growth	leafy bud developed from protonema, 0.1 mm in size	Plants 3-4 leaf stage, about 0.30 mm in size	-	-

(Table 1) cont.....

Days	Culture Media					
	½ Knop's 0.1 mg/L IBA+ 0.2 mg/L BAP (A)	½ Knop's 0.2 mg/L IBA+0.2 mg/L BAP (B)	½ Knop's 0.1 mg/L 2,4D+0.1 mg/L Kinetin (C)	½ Knop's 0.1 mg/L IAA +0.1 mg/L BAP (D)	Hoagland	½ Knop's
(68th day)	Death of Protonema	Death of protonema	Plants about 1 mm in size, dense brown rhizoids developed at the base of the plant.	Plants about 0.4 to 0.6 mm in size, dense brown rhizoids developed at the base of the plant.	-	-

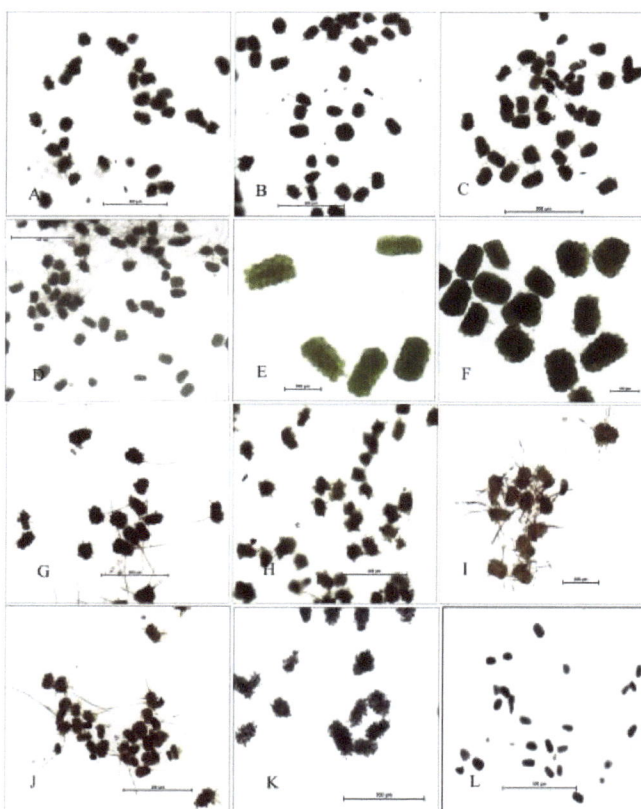

Plate (2). *In-vitro* growth and multiplication of *Drummondia stricta*, A-F: Growth of spores in different media after 6 days, A. In medium A, B. In medium B, C. In medium C, D. In medium D, E. In Hoagland medium, F. In ½ Knop's medium; G-L: Growth of Spores in different media after 15 days, G. In medium A, H. In medium B, I. In medium C, J. In medium D, K. In Hoagland medium, L. In ½ Knop's medium.

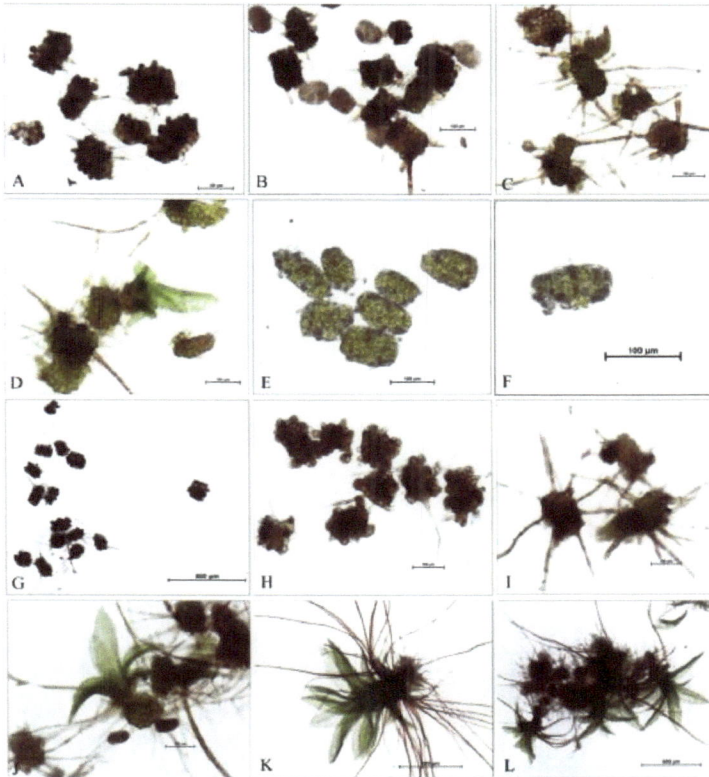

Plate (3). *In-vitro* growth and multiplication of *D. stricta*. Protonema development in different media after 30 days, A-F: A. In medium A, B. In medium B, C. In medium C, D. In medium D, E. In Hoagland medium F. In ½ Knop's medium; G-J: Protonema and leafy bud development in different media after 45 days, G. In medium A, H. In medium B, I. In medium C, J. In medium D; K-L: Leafy gametophores development after 68 days, K. In medium C, L. In medium D.

CONSENT FOR PUBLICATION

Not applicable.

CONFLICT OF INTEREST

The authors confirm that this chapter contents have no conflict of interest.

ACKNOWLEDGEMENTS

Authors are thankful to the Director, CSIR-National Botanical Research Institute, Lucknow, India for encouraging and providing the facilities.

REFERENCES

[1] Vitt DH. A monograph of the genus *Drummondia*. Can J Bot 1972; 50: 1191-208.
[http://dx.doi.org/10.1139/b72-145]

[2] Chopra RS. Taxonomy of Indian Mosses. New Delhi: C.S.I.R. Publication 1975; pp. 1-691.

[3] Kaul KN, Mitra GC, Tripathi BK. Morphogenetic responses of the thallus of *Marchantia* to several growth substances. Curr Sci 1961; 4: 131-33.

[4] Kaul KN, Mitra GC, Tripathi BK. Responses of *Marchantia* in aseptic culture to well known Auxins and Antiauxins. Ann Bot 1962; 26: 447-66.
[http://dx.doi.org/10.1093/oxfordjournals.aob.a083806]

[5] Asthana AK, Sahu V. Growth responses of a moss *Brachymenium capitulatum* (Mitt.) Par. in different culture media. Natl Acad Sci Lett 2011; 34(1-2): 1-4.

[6] Awasthi V, Nath V, Asthana AK. Effect of some physical factors on reproductive behaviour of selected bryophytes. Int J Plant Reprod Biol 2010; 2(2): 141-45.

[7] Sahu V, Asthana AK. An observation on growth response of *anomobryum filiforme* var. *concinnatum* (Spruce) aman. (Bryaceae) in different culture media. Natl Acad Sci Lett 2013; 36(6): 587-89.
[http://dx.doi.org/10.1007/s40009-013-0173-8]

[8] Asthana AK, Sahu V, Srivastava A. *In-vitro* propagation of three species of *Bryum* Hedw.: A comparative study. Geophytology 2015; 45(2): 215-20.

[9] Srivastava A, Sahu V, Asthana AK. An observation on morphogenetic response of *marchantia polymorpha* subsp. *ruderalis* bischl. & boissel. dub. in different culture media. International Journal of Plant and Environment 2018; 4(1): 64-9.

[10] Sahu V, Asthana AK. A study on photomorphogenesis of protonema and bud formation in *pohlia ludwigii* (spreng. ex schwägr.). broth. Indian forester 2018; 144(8): 781-3.

[11] Nehira K. Spores germination, Protonema development and sporeling development New Manual of Bryology. Schuster RM. Japan: Miyazaki, The Hattori Botanical Laboratory, Nichinan, Miyazaki 1983; pp. 343-85.

[12] Bopp M, Atzorn R. Hormonelle regulation der moosentwicklung. Naturwissenschaften 1992; 79: 337-46.
[http://dx.doi.org/10.1007/BF01140176]

[13] Gao YC, Sha W, Zhang H. Effect of different plant growth substances on callus induction of *Cratoneuron filicinum*. Plant Physiology communications 2003; 39: 29-31.

[14] Chen JW, Cao T, Shi DJ. The effect of plant hormones to calli inducement and differentiation of *Physcomitrella patens* (Hedw.) B.S.G. Journal of Shanghai Normal University 2006; 35: 70-4. [Natural Sciences].

[15] Cvetic T, Sabovljevic A, Sabovljevic M, Grubisic D. Development of the moss *Pogonatum urnigerum* (Hedw.) P. (DH 1972) Beauv. under *in-vitro* culture conditions. Arch Biol Sci 2007; 59: 57-61.
[http://dx.doi.org/10.2298/ABS0701057C]

CHAPTER 2

Bryophytes and their Associates in South Shetland Islands -Antarctica

Jair Putzke[*]

University UNIPAMPA, Av. Antonio Trilha, 1847 - São Clemente, São Gabriel - RS, CEP: 97300-000 - Brazil

Abstract: Bryophytes constitute one of the most prominent groups among terrestrial vegetation in South Shetlands Islands of Antarctica. The present study is focused on the bryoflora which exists in association with other life forms in the studied regions. Though, the taxonomy of the Antarctic bryophytes has received much attention in recent years but their associations, especially with lichens and fungi/algae are somewhat neglected by earlier explorers. However, these associations are very crucial for plant succession in the area like Antarctica. A checklist of bryophytes has been provided along with their associated forms with short discussion on the associations to signify the importance of these tiny plants.

Keywords: Antarctica, Associations, Bryophytes, Lichen, Succession.

INTRODUCTION

Antarctica (Plate **1**; Fig. **1**) was the last continent discovered around 1820 as one of the most fascinating region on the earth with characteristic flora and fauna. At present this remote area of our planet is one of the most drastically changing zones. Climate change is another concern which directly affecting the Antarctic ecosystem in a negative way, affecting the overall quality and quantity of Antarctic species.

Floristic research on Antarctic has received much attention in recent years, especially in context to the influence of climate change and ozone layer depletion on the flora. However, the major research was invariably related to the taxonomy of various plant groups through basic surveys.

The main area of study for bryologists in Antarctica has been the South Shetland Islands (Plate **1**; Fig. **2**), an archipelago of at least 8 larger islands and dozens of

---

[*] **Corresponding author Jair Putzke:** University UNIPAMPA, Av. Antonio Trilha, 1847 - São Clemente, São Gabriel - RS, CEP: 97300-000 - Brazil; Tel: 51 37177518; E-mail: jrputzkebr@yahoo.com

Afroz Alam (Ed.)

smaller, where most of the scientific bases are installed and doing research. However, regarding bryophytes, most of the information has been generated in surveys of these islands that resulted in several checklists of Antarctic bryoflora.

Plate 1. Figure 1. General view of Antarctica; Figure 2. General view of the South Shetland Islands – Elephant Island (Bryorich region).

Ecologically the bryophytes of the Antarctic constitute one of the main components of ecosystem among all the terrestrial organisms present on this continent (Plate **2**; Fig. **3**). They have important relations with several organisms, being able to be associated with lichens, fungi, algae, higher plants, invertebrates and terrestrial vertebrates. The varied associations denote the importance of these tiny organisms of plant kingdom for the ecology of the studied area.

Plate 2. Figure 3: Luxuriant growth of terrestrial bryophytes; Figure 4: Sparse growth of bryophytes along with *Marchantia*; Figure 5: Growth of *Cephaloziella* in isolated patches; Figure 6: Patch of Pleurocarpous moss *Brachythecium*; Figures 7 & 8: Patch of acrocarpic moss.

In present work an attempt has been made to collect all inclusive information regarding those bryophytes only that are living with associations of other groups, especially lichens, fungi and algae. A table is provided containing bryophytes their life forms and associates forms.

Prevalent Bryophytes

Among the bryophytes, mosses (Bryopsida) are the most prevalent that occupy major terrestrial areas including bare rocks. At places few patches of liverworts (Marchantiophyta) are also found. Genus *Marchantia*, a thalloid liverwort also found but without any main association. Leafy liverworts come next to the mosses and have better frequency than the thalloid liverworts. Among the leafy liverworts, *Cephaloziella varians* was found associated with a mycorrhizal fungus *Rhizoscyphus ericae* (ericoid symbiosis) throughout Antarctica [1]. Occurrence of *Cephaloziella varians* explains the importance of fungal association that allows such fragile species to conquer this extreme environment (Plate **2**: Figs. **4**, **5**). It also reflects an interdependence of the associated species and, if one species is threatened, the other will also be threatened [2].

The vitality of these associations is further strengthened by one of the oldest fossil records in the form of *Schizolepidella gracilis* Halle, member of Marchantiophyta (Jungermanniales) which was found in the Antarctic Peninsula along with lichens from the Jurassic remains [3], a period when Antarctica was assumed as a tropical continent.

Till date, about 112 species, with 55 genera under 17 families have been reported to Antarctica [1, 4]. These species have been categorized in two main growth forms: the pleurocarpic (Plate **2**: Fig. **6**), where the stalk of the plant is prostrate, forming continuous carpets and acrocarpic (Plate **2**: Fig. **7, 8**), where plants grow upright, erect, forming tufts or smaller cushions. Remarkably, among all species of bryophytes, the moss *Sanionia uncinata* with its characteristic curved leaves, shows highest occurrence and biomass at all accessible localities as a carpet forming taxa.

It is an interesting point that due to extremely low temperature environment of Antarctica, the bryophytes can stay viable for many years, *e.g.*, in 2014, a survey showed that the moss *Chorisodontium acyphyllum* (Plate **3**: Fig. **9**) remained alive after held frozen for over 1500 years. To further confirm that, a 1.4-meter-thick turf was sectioned 20-by-20 cm (layer per layer) and placed to germinate under ideal conditions. In 3-8 weeks all restarted the growth. The deepest layer was radiocarbon dated and estimated between years 1533 and 1697 [5].

Plate 3. Figure 9: Association of terrestrial bryophytes with lichens; Figure 10: Cushion forming bryophytes; Figure 11: Different zones of an elevated landscape (A - nitrophobous community; B- nitrophilous community; C-halophyte community; Figure 12: Growth of *Pohlia cruda* and *Bartramia patens* under nitrogen limiting environment, *e.g.*; Figures 13: Patchs of *Bryum pallescens*.

The most available substratum in Antarctica are rocks on which moss taxa *viz.*, *Andreaea* (Plate **3**: Fig. **10**) and *Schistidium*, species that are found in association with crustose lichens exclusively as saxicolous forms [3].

Since the study areas are islands therefore halophytic associations are also existing and such bryophytes prefer to colonize in areas closest to the sea or at least receiving splashes of saltwater waves. The species *Muelleriella crassifolia* P. Dusén requires at least some contact with sea salt to develop. Eventually this species was observed with lichen genus *Verrucaria*, another halophytic species [6] (Plate **3**: Fig. **11**).

Associations can be observed with the native phanerogams of the continent and among mosses and lichenized fungi and Marchantiophyta. There is also some preference for nutrient availability, especially for the presence or proximity to nesting, resting or birding sites. These species are called ornitocoprophilous or nitrophilous and they grow many times in slopes bathed by the excrements of the animals that occur above. Moss taxa, such as *Synchitria magellanica* and *Henediella heimii* (Plate **3**: Fig. **9**) prefer to grow in association of native phanerogams along with lichenized fungi [7].

Another group of bryophytes is ornithopophobic or nitrophobic, represented by those species that prefer to grow in nitrogen limiting environment, *e.g. Pohlia cruda* and *Bartramia patens* (Plate **3**: Fig. **12**). These taxa were found associated with lichens that can provide support in this limiting environmental restrain.

Major Associations of Bryophytes

The all inclusive associations most commonly found in the South Shetland Islands of Antarctica can be categorized as following [8]:

Association with Fruticose Lichens

This association consists basically of lichens such as *Usnea aurantiacoatra*, *Cladonia* spp., *Stereocaulon glabrum* and *Sphaerophorus globosus* and mosses *viz.*, *Sanionia* spp., and *Polytrichastrum alpinum*. Further, this association can alternate with others in a gradual way, including in cases where the saxicolous lichens are more represented, when *Usnea aurantiacoatra* is the most found (Table **1**).

Association Between Carpet Forming Mosses and Lichens

This community is formed between the moss *Sanionia* spp. having the highest biomass in the study area. The carpets were found infested with lichens *viz.*, *Stereocaulon glabrum*, *Sphaerophorus globosus* and *Psoroma hypnorum*, *Ochrolechia frigida* and *Cladonia* spp. (Table **1**), *etc.*

Association of Bryophytes with Muscicolous Lichens

This association may be the most expressive in the ice-free areas of some islands. About 29 species were found in this association. Though, the moss taxa *Sanionia* spp. is mostly found associated with muscicolous lichens but other bryophytes are also reported in this type of association depending on the drainage of the terrain. Bryophytes usually colonized by crustose, foliose and fruticose lichens, *viz.*, *Cladonia* spp., *Ochrolechia frigida*, *Psoroma hypnorum*, *Sphaerophorus globosus* and *Stereocaulon glabrum* [9] (Table **1**).

Table 1. Diversity of most common Bryophytes in South Shetland in Antarctica and their associations in which the species is found and associated species in each association.

S. No.	Name of the Bryophyte	Associations in which the Species are Found	Associated Species
1.	*Andreaea depressinervis* Card.	Saxicolous	*Verrucaria* spp. [lichenized (lichen-forming) fungi] *Cladonia metacorallifera* Asahina (Lichen)
		Turf forming	*Buellia anisomera* Vain. (lichen) *Buellia latemarginata* Darb. (lichen)
		Carpet forming	*Cladonia borealis* S. Stenroos *Himantormia lugubris* (Hue) Cordeiro IM
		Muscicolous	*Cladonia rangiferina* (L.) Weber ex F.H. Wigg. *Cladonia metacorallifera* Asahina
2.	*Andreaea gainii* Card.	Saxicolous	*Lecidea* sp. (crustose lichen)
		Carpet forming	*Rhizocarpon polycarpum* (Hepp) Th. Fr. *Sphaerophorus globosus* (Huds.) Vain. *Rhizoplaca aspidophora* (Vain.) Redon
		Muscicolous	*Rhizoplaca aspidophora* (Vain.) Redon *Cladonia borealis* S. Stenroos
		Crustose	*Acarospora macrocyclos* Vain.
3.	*Andreaea regularis* Muell.	saxicolous	*Acarospora macrocyclos* Vain. (lichen) *Rhizoplaca melanophthalma* (Ram.) Leuckert &Poelt (lichen)

(Table 1) cont.....

S. No.	Name of the Bryophyte	Associations in which the Species are Found	Associated Species
		Turf forming	*Stereocaulon glabrum* (Müll. Arg.) Vain *Rhizoplaca aspidomorpha* (Vain.) Radeon
		Carpet forming	*Verrucaria* sp. Schrad *Psoroma hypnorum* (Vahl) Gray *Prasiola crispa*(Lightfoot) Kützing (terrestrial green algae)
		Muscicolous	*Psoroma hypnorum* (Vahl) Gray *Stereocaulon glabrum* (Müll. Arg.) Vain
		Crustose	*Turgidosculum complicatulum* (Nyl.) Kohlm. & E. Kohlm. (fungus) *Ochrolechia frigida* (Sw.) Lynge
4.	*Bartramia patens* Brid.	Turf forming	*Prasiola crispa* (Lightfoot) Kützing (algae) *Poroma hypnorum* (Vahl) Gray (lichen)
		Muscicolous	*Caloplaca athallina* Darb. *Cladonia furcata* (Huds.) Schrad.
		Carpet forming	*Pannaria hookeri* (Borrer) Nyl. *Psoroma cinnamomeum* Malme
		Crustose	*Buellia latemarginata* Darb.
5.	*Brachythecium austrosalebrosum* (Müll. Hal.) Kindb	Carpet forming	*Caloplaca athallina* Darb
6.	*Bryum argenteum* Hedw.	Turf forming	*Cladonia rangiferina* (L.) Weber *ex* F.H. Wigg. (lichen) *Lecidea sciatrapha* Hue (lichen)
		Carpet forming	*Acarospora macrocyclos* Vain. *Usnea antarctica* Du Rietz
7.	*Bryum dichotomum* Hedw.	Carpet forming	*Haematomma erythromma* (*Nyl.*) *Zahlbr.*
8.	*Bryum nivale* Müll. Hal	Carpet forming	*Huea austroshetlanica* (Zahlbr.) C.W. Dodge
9.	*Bryum niveum* Herzog	Carpet forming	*Cladonia rangiferina* (L.) Weber ex F.H. Wigg *Cladonia metacorallifera* Asahina
10.	*Bryum pallescens* Schleich. *ex* Schwägr. (Plate **3**: Fig. **13**)	Carpet forming	*Microglaena antarctica* IM Lamb
11.	*Bryum pseudotriquetrum* (Hedw.) P.Gaertn., E. Meyer & Scherb.	Crustose	*Bacidia tuberculata* Darb. *Usnea aurantiacoatra* (Jacq.) Bory
12.	*Bryum pseudotriquetrum* (Hedw.) Schwaegr.	Carpet forming	*Bacidia tuberculata* Darb. *Buellia russa* (Hue) Darb.

(Table 1) cont.....

S. No.	Name of the Bryophyte	Associations in which the Species are Found	Associated Species
13.	*Cephallozia* sp. (Dumort.) Dumort.	Saxicolous	*Ochrolechia parella* (L.) A. Massal. *Parmelia saxatilis* (L.) Ach.
14.	*Ceratodon grossiretis* Cardot	Carpet forming	*Lecanora skottsbergii* Darb.
15.	*Ceratodon purpureus* (Hedw.) Brid.	Turf forming	*Psoroma cinnamomeum* Malme (lichenized fungi) *Usnea aurantiacoatra* (Jacq.) Bory (lichen)
		Saxicolous	*Psoroma cinammomeum* Malme. *Rhizocarpon geographicum* (L.) DC.
		Crustose	*Caloplaca regalis* (Vain.) Zahlbr. *Xanthoria candelaria* (L.) Th. Fr.
16.	*Chorisodontium oxyphyllum* (Hook f. *et.* Wills.) Broth.	Turf forming	*Rhizocarpon geographicum* (L.) DC. (lichen) *Rhizocarpon polycarpum* (Hepp) Th. Fr. (lichen)
17.	*Dicranoweisia grimmiacea* (Müll. Hal.) Broth.	Saxicolous	*Rhizocarpon polycarpum* (Hepp) Th. Pe
18.	*Hennediella antarctica* (Angstr.) Ochyra &Matteri	Crustose	*Haematomma erythromma* (Nyl.) Zahlbr. *Pannaria hookeri* (Borrer) Nyl.
19.	*Hennediella heimii* (Hedw.) Zand. (Heim's pottia moss)	Carpet forming	*Stereocaulon glabrum (*Müll. Arg.) Vain. *Cystocoleus niger (*Huds.) Har.
		Muscicolous	*Microglaena antarctica* IM Lamb *Prasiola crispa* (Lightfoot) Kützing (alga)
		Turf forming	*Verrucaria* sp. (lichenized fungi)
20.	*Himantormia lugubris* (Hue) Cordeiro IM	Cushion forming	*Leptogium puberulum* Hue
		Muscicolous	*Acarospora macrocyclos* Vain. *Psoroma cinnamomeum* Malme
21.	*Pohlia cruda* (Hedw.) Lindb.	Turf forming	*Cystocoleus niger* (Huds.) Har. (fungi)
		Carpet forming	*Buellia anisomera* Vain. *Buellia racovitzae* CW Rodeio
		Muscicolous	*Usnea antarctica* Du Rietz *Rhizocarpon geographicum* (L.) DC. *Buellia latemarginata* Darb.
22.	*Pohlia nutans* (Hedw.) Lindb.	Carpet forming	*Ochrolechia frigida* (Sw.) Lynge
		Muscicolous	*Rhizoplaca melanophtalma* (Ram.) Leuckert & Poelt *Huea austroshetlandica* (Zahlbr.) CW Dodge

(Table 1) cont.....

S. No.	Name of the Bryophyte	Associations in which the Species are Found	Associated Species
23.	*Polytrichastrum alpinum* G.L. Smith	Corticolous	*Ochrolechia frigida* (Sw.) Lynge (lichen) *Lecidea sciatrapha* Hue (Lichen)
		Carpet forming	*Usnea aurantiacoatra* (Jacq.) Bory *Ochrolechia frigida* (Sw.) Lynge
		Muscicolous	*Sphaerophorus globosus* (Huds.) Vain. *Rhizocarpon polycarpum* (Hepp) Th. Fr.
		Turf forming	*Ochrolechia frigida* (Sw.) Lynge (Lichen) *Sphaerophorus globosus* (Huds.) Vain. (lichen)
24.	*Polytrichum piliferum* Hedw. (Bristly Haircap moss)	Saxicolous	*Cladonia rangiferina* (L.) Weber ex F.H. Wigg. *Buellia latemarginata* Darb. *Cystocoleus niger* (Huds.) Har.
		Carpet forming	*Rhizoplaca melanophtalma* (Ram.) Leuckert & Poelt *Buellia latemarginata* Darb.
		Turf forming	*Cladonia metacorallifera* Asahina (lichen)
25.	*Sanionia uncinata* (Hedw.) Loeske (Sickle-leaved Hook-moss)	Saxicolous	*Usnea aurantiacoatra* (Jacq.) Bory (lichen) *Sphaerophorus globosus* (Huds.) Vain (lichen)
		Carpet forming	*Usnea aurantiacoatra* (Jacq.) Bory *Lecidea sciatrapha* Hue
		Muscicolous	*Usnea aurantiacoatra* (Jacq.) Bory *Lecidea sciatrapha* Hue
		Crustose	*Lecania brialmontii* (Vain.) Zahlbr.
		Turf forming	*Rhizoplaca aspidophora* (Vain.) Redon (lichen)
26.	*Schistidium antarctici* (Card.) L. I. Savic. & Smirn	Crustose	*Candelaria murrayi* Poelt *Lecidea sciatrapha* Hue
		Muscicolous	*Physconia muscigena* (Ach.) Poelt *Umbilicaria antarctica* Frey & IM Lamb
27.	*Schistidium steerii* Ochyra	Saxicolous	*Microglaena antarctica* IM Lamb
28.	*Schistidium urnulaceum* (Müll. Hal.) BG Sino	Saxicolous	*Buellia russa* (Hue) Darb. *Caloplaca cinericola* (Hue) Darb.
29.	*Syntrichia magellanica* (Mont.) RH Zander	Crustose	*Rinodina petermannii* (Hue) Darb.

Association of Bryophytes with Crustose Lichens

This association is chiefly composed of crustose lichens with moss genera like *Andreaea* spp. and *Schistidium* spp. (Table **1**).

Association of Turf Mosses with Lichens

Erect-growing mosses of the family Polytrichaceae found most common in some areas where they form peat bogs. Usually these mosses prefer to grow alone however, during protonemal stage they are found in association of lichens [10, 11] (Table **1**).

DISCUSSION

On the basis of this observation regarding the associations of these pioneer plant groups it is revealed that the harsh environment of the Antarctic region has stimulated these associations that are necessary for the sustenance of bryophytes and the lichens, both [12 - 14]. The intimate associations are imperative to counter the environmental conditions of the region and somewhat constitute the backbone of the ecosystem by pioneering the plant succession in such a harsh climatic zone of the world. Ecologically, these associations share the same ecological niche with different basic requirements to grow that are complementary to each other. Out of 112 reported bryophytes, 29 species have shown to make part of these five main associations which indicates that these species are fragile when grow alone. In case of habitat range moss taxa, *Andreaea* spp. and *Sanionia uncinata* showed maximum associations diversity followed by *Polytrichastrum alpinum purpureus* which means that these taxa have greater adaptations than the other. The outcome revealed that there is a great need of conservation methods of these neglected but imperative plants in Antarctica.

CONSENT FOR PUBLICATION

Not applicable.

CONFLICT OF INTEREST

The authors confirm that this chapter contents have no conflict of interest.

ACKNOWLEDGEMENTS

The author is grateful to the University of UNIPAMPA for encouragement and support.

REFERENCES

[1]   Newsham KK. Structural changes to a mycothallus along a latitudinal transect through the maritime and sub-Antarctic. Mycorrhiza 2011; 21(3): 231-6.
[http://dx.doi.org/10.1007/s00572-010-0328-0] [PMID: 20628887]

[2]   Lewis-Smith RI. Plant colonization response to climate changes in the Antarctica. Folia Fac. Sci. nat. univ. Masarykianae Brunensis. Geográfica 2001; 25: 19-33.

[3]   Ociepa AM. Jurassic liverworts from Mount Flora, Hope Bay, Antarctic Peninsula. Pol Polar Res 2007; 28(1): 31-6.

[4]   Ochyra R, Lewis-Smith RI, Bednarek-Ochyra H. The illustrated moss flora of Antarctica. Cambridge University Press 2008.

[5]   Roads E, Longton RE, Convey P. Millennial timescale regeneration in a moss from Antarctica. Curr Biol 2014; 24(6): R222-3.
[http://dx.doi.org/10.1016/j.cub.2014.01.053] [PMID: 24650904]

[6]   Pereira AB, Putzke J. Floristic Composition of Stinker Point, Elephant Island, Antarctica. Korean J Polar Res 1994; 5(2): 37-47.

[7]   Albuquerque MP, Victoria FC, Schünemann AL, *et al.* Plant Composition of Skuas Nests at Hennequin Point, King George Island, Antarctica. Am J Plant Sci 2012; 3: 688-92.
[http://dx.doi.org/10.4236/ajps.2012.35082]

[8]   Casanova-Katny MA, Cavieres LA. Asntarctic moss carpets facilitate growth of *Deschampsia antarctica* but not its survival. Polar Biol 2012; 35: 1869-78.
[http://dx.doi.org/10.1007/s00300-012-1229-9]

[9]   Onofre S, Zucconi L, Tosi S. Continental Antarctic Fungi. Eching Bei München : Ihw-Verlang. 2007.

[10]   Putzke J, Pereira AB. Fungos muscícolas na Ilha Elefante, antártica. Cad Pesqui Sér Biol 2012; 21(1): 155-64.

[11]   Putzke J, Pereira AB. Macroscopic fungi from the South Shetlands, Antarctica. Serie Cientifica INACH 1996; 46: 31-9.

[12]   Gumińska B, Heinrich Z, Olech M. Macromycetes of the South Shetland Islands (Antarctica). Pol Polar Res 1994; 15(3-4): 103-9.

[13]   Lewis LR, Behling E, Gousse H, *et al.* First evidence of bryophyte diaspores in the plumage of transequatorial migrant birds. PeerJ 2014; 2: 1-13.
[http://dx.doi.org/10.7717/peerj.424]

[14]   Putzke J, Pereira AB, Victoria FC, Pereira CK, Dóliveira CB, Schünemann AL. Plant Communities from Stinker Point, Elephant Island - Antarctica. Annual Activity Report of National Institute of Science and Technology Antarctic Environmental Research 2012; 1: 54-7.

CHAPTER 3

Computational Resources for Bryology

Sonu Kumar and **Asheesh Shanker**[*]

Department of Bioinformatics, Central University of South Bihar, Gaya- 824236, India

Abstract: Computational/Bioinformatics resources play an important role in biological research, including studies on bryophytes, a group of non-vascular land plants categorized into hornworts, liverworts, and mosses. These resources deal with the development and application of computational approaches to solve biological problems. The availability of such resources, including databases, software, and other tools are useful in the analysis and interpretation of biological data. This chapter provides a description of various useful bioinformatics resources, which can bring a major change in terms of time, money, and labor in studies related to bryophytes.

Keywords: Bioinformatics, Bryophytes, Databases, Software.

INTRODUCTION

Bryophytes are the earliest land plants that are categorized into hornworts (Anthocerotophyta), liverworts (Marchantiophyta), and mosses (Bryophyta) [1]. These land plants are found in almost all ecosystems and play significant role in environment functioning, global biogeochemical cycles, and also influence vegetation dynamics [2]. Moreover, these small and simple non-vascular plants are used as a model to study the complex biological mechanisms in plant research [3].

With the recent technological progress, an unprecedented development has been observed in plant science, which provided a large amount of data. The analysis of such data available in the public databases is a challenge for researchers from different biological fields. However, development of computational resources, including databases, software and web servers can result in a major change in data analysis. Recently, the use of various bioinformatics resources in the study of plant stress [4] and databases for medicinal plant research [5] were described. Data emerging from bryophytes research also require suitable tools for management and analysis.

[*] **Corresponding Author Asheesh Shanker:** Department of Bioinformatics, Central University of South Bihar, Gaya-824236, India; Tel: +91-9414478655; E-mail: ashomics@gmail.com

Afroz Alam (Ed.)

Previously, various studies have been reported on bryophytes including microsatellites identification [6 - 11], phylogeny [12 - 14], protein-protein interactions [15], and variation in secondary metabolites [16] using different computational approaches.

Computational/ Bioinformatics Resources

Considering the importance of bioinformatics in the field of plant science, different computational approaches, including tools, databases, and web resources are described here that can be helpful in studies related to bryophytes.

Sequence Analysis

Biological sequences, including deoxyribonucleic acid (DNA), ribonucleic acid (RNA), and protein (amino acid) sequences are one of the most fundamental objects to study biological system at the molecular level. The easiest way to assess the relationship between two or more species is to identify the identical and similar pattern of the biological sequence between them. This can be done with the help of sequence alignment, which is one of the most important bioinformatics technique widely applied to identify the similarity between these sequences [17, 18].

To date several computational tools have been developed to perform sequence alignment and such algorithms are broadly divided into local [19] and global [20] alignment algorithms. Local alignment algorithms are used to identify only the most similar pattern within sequences, whereas global alignment algorithms align the whole sequence.

Basic Local Alignment Search Tool (BLAST; http://ncbi.nlm.nih.gov/BLAST/) [21, 22] is one of the most common tools used to align nucleotides, and amino acid sequences. BLAST applies a heuristic method to find local similarity and calculates the statistical significance of matches to infer functional and evolutionary relationships between sequences. FASTA (https://www.ebi.ac.uk/Tools/sss/fasta/) [23] is also a sequence alignment tool based on the heuristic search algorithm to find significant matches between biological sequences. Both BLAST and FASTA are useful to align sequences in large-scale databases. Moreover, Multiple Sequence Alignment (MSA) is also a broadly applied method, which is used to align more than two sequences to infer homology and the evolutionary relationship among sequences [24]. ClustalW (http://ebi.ac.uk/Tools/msa/clustalw2/) [25], a fully automated program for global multiple sequence alignment of DNA and protein sequences is available. Apart from this, T-COFFEE (http://tcoffee.crg.cat) [26], MUSCLE (http://ebi.ac.uk/Tools/msa/muscle/) [27], MAFFT (http://mafft.cbrc.jp/alignment/ software/) [28], and DIALIGN2 (http://bibiserv.techfak.uni-bielefeld.de/dialign/) [29] have also been developed to perform multiple sequence alignments.

Earlier, MSA methods have been applied for the analyses of nad5 gene sequences in 47 bryophytes [30], Polytrichales [31], and using organellar genome sequences [12 - 14, 32]. The alignment tools discussed here will play an important role in investigating homology and phylogenetic relationship in bryophytes. Moreover, these tools are useful to infer the function of newly sequenced genes by aligning them with known gene sequences submitted in databases. Sequence alignments can also play a vital role in the template identification during protein structure prediction.

Structure Prediction

Structure prediction is one of the widely used techniques, which incorporate computational approach to build three-dimensional (3D) structures of macromolecules. Further, it is categorized into three different forms, namely Homology or comparative modelling, Threading or fold recognition, and *ab initio* method.

Homology modelling has been used to predict a 3D structure of a protein from its sequence alone by identifying homologous sequence whose structure is known [33]. Threading approach has been used to model proteins with very low sequence similarity with other proteins, however, showing good compatibility with a known fold [34]. Moreover, *ab initio* method has been used to build a model structure in the absence of suitable homologous sequence/structure.

A number of tools have been developed to build 3D structure of proteins. SWISS-MODEL (http://swissmodel.expasy.org/) is an online automated server developed for homology modelling [35]. MODELLER (http://www.salilab.org/modeller/) is widely used program for homology modeling of 3D structures of protein based on known template structure [36]. ESyPred3D (http://www.unamur.be/sciences /biologie/urbm/bioinfo/esypred/) [37], ROBETTA (http://robetta.bakerlab.org/) [38], Bhageerath (http://www.scfbio-iitd.res.in/bhageerath/index.jsp/) [39], Local Meta-Threading-Server (LOMETS; http://zhanglab.ccmb.med.umich.edu/ LOMETS/) [40], CPHmodels (http://www.cbs.dtu.dk/services/CPHmodels/) [41], CABS-fold (http://biocomp.chem.uw.edu.pl/CABSfold/) [42], RaptorX (http://raptorx.uchicago.edu/) [43], PEP-FOLD (http://bioserv.rpbs.univ-paris-diderot.fr/services/PEP-FOLD/) [44], Iterative Threading ASSEmbly Refinement (I-TASSER; http://zhanglab.ccmb.med.umich.edu/I-TASSER/) [45] *etc.*, have also been developed to model 3D structure of protein.

In addition to this, tools are also available to assess the quality of the modelled protein structure. ProTSAV (http://www.scfbio-iitd.res.in/software/proteomics /protsav.jsp) is one of the useful resources to evaluate the quality of the predicted 3D model of a protein [46].

The molecular modelling tools discussed here will be useful to predict the 3D structure of proteins using amino acid sequence. The predicted structure can further be used for functional analysis.

Molecular Docking and Molecular Dynamic Simulation

Docking is a widely used method to analyze interaction between molecules. It helps to predict the binding affinity of a small molecule (ligand) to their target protein. Till now, many computational tools have been developed to study the interaction between molecules. These tools can also be used in the field of bryology.

AutoDock (http://autodock.scripps.edu/) [47] is one of the widely applied docking tools designed to predict interactions between ligand and receptor. AutoDock Vina (http://vina.scripps.edu/) [48] is a new generation of docking software, which is faster than AutoDock and achieves significant perfections in the prediction of binding mode of ligand. Apart from this other well established tools for protein-ligand docking include Glide (http://www.schrodinger.com/glide) [49], Generic Evolutionary Method for molecular DOCKing (GEMDOCK; http://gemdock.life.nctu.edu.tw/dock/) [50], and SANJEEVINI (http://www.scfbio-iitd.res.in/sanjeevini/sanjeevini.jsp/) [51].

Molecular dynamics (MD) simulation methods can be used to study structural changes of a molecule with respect to time and to refine the modelled structure. Various tools are available, including Assisted Model Building and Energy Refinement (AMBER; http://ambermd.org/) [52], Chemistry at Harvard Macromolecular Mechanics (CHARMM; https://www.charmm.org/) [53], and GROningen MAchine for Chemical Simulations (GROMACS; http://www.gromacs.org/) [54] to perform MD simulation.

These molecular docking and MD simulation tools will be useful to analyze protein-ligand interactions and structural changes observed in proteins with respect to time, respectively. Moreover, these tools will also play an important role in computer aided drug discovery using chemical compounds extracted from bryophyte species.

Despite the availability of various protein modelling, molecular docking, and MD simulation tools there is a scarcity of such studies for bryophytes.

Database Resources

Biological databases are collection of biological information that store, organize, and share biological data [55]. Based on data coverage these databases are classified as comprehensive and specialized databases. Comprehensive databases contain various types of data from numerous sources, whereas specialized databases contain specific data or data from specific organisms/species.

Primary database, including GenBank (http://www.ncbi.nlm.nih.gov/genbank/) [56], European Molecular Biology Laboratory's European Bioinformatics Institute (EMBL-EBI; http://www.ebi.ac.uk/embl/), and DNA Data Bank of Japan (DDBJ; http://www.ddbj.nig.ac.jp/), which are part of International Nucleotide Sequence Database Collaboration (INSDC; http://www.insdc.org) contains DNA and RNA sequences from different organisms including bryophytes.

Moreover, Protein Information Resource (PIR; http://pir.georgetown.edu/) contains functionally annotated protein sequences [57], NCBI Protein database (https://www.ncbi.nlm.nih.gov/protein/), a collection of protein sequences from numerous sources [58], UniProtKB/Swiss-Prot (http://www.uniprot.org/uniprot/) database make available protein sequences along with their functional information [59], Protein Data Bank (PDB; http://www.rcsb.org/pdb/home/home.do/) provides the 3D structures of proteins, nucleic acids, and complex assemblies [60], The Molecular Modeling Database (MMDB; http://www.ncbi.nlm.nih.gov/Structure/MMDB/mmdb.shtml/) contains 3D biomolecular structures [61], The Kyoto Encyclopedia of Genes and Genomes (KEGG; https://www.genome.jp/kegg/) provides information about biological pathways, genomes, chemical substances, diseases, and drugs [62], EnsemblPlants (http://plants.ensembl.org/index.html/) is a genome centric portal for plant species [63], and PubMed (https://www.ncbi.nlm.nih.gov/pubmed/) which provides a catalog of scientific literature are few examples of biological databases to deliver knowledge and information of biology to scientific community. Select resources which contain scientific information about bryophyte species are presented here.

The Plant List

The Plant List (http://www.theplantlist.org/; Fig. **1**) database includes a list of all known plant species including bryophytes. The database contains around 60805 scientific names of bryophyte species out of which 34556 are accepted. These bryophyte species belong to 177 plant families and 1822 plant genera. Moreover, The Plant List also contains information about pteridophytes, gymnosperms, and angiosperms.

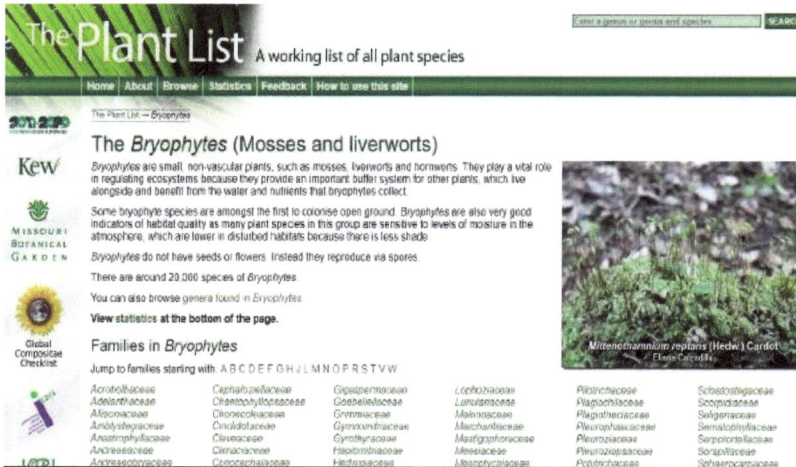

Fig. (1). The Plant List database.

Plant Transcription Factor Database

The Plant Transcription Factor Database (PlnTFDB; http://plntfdb.bio. uni-potsdam.de/v3.0/; Fig. **2**) contains information about all Plant genes, including bryophytes, involved in transcriptional control. The database provides protein models (1305) and distinct protein sequences (1295) of *Physcomitrella patens*, organized in 72 gene families. Moreover, a total of 118 proteins was categorized as orphans and classification of these proteins into any defined families and role in the transcriptional regulation remains unclear [64].

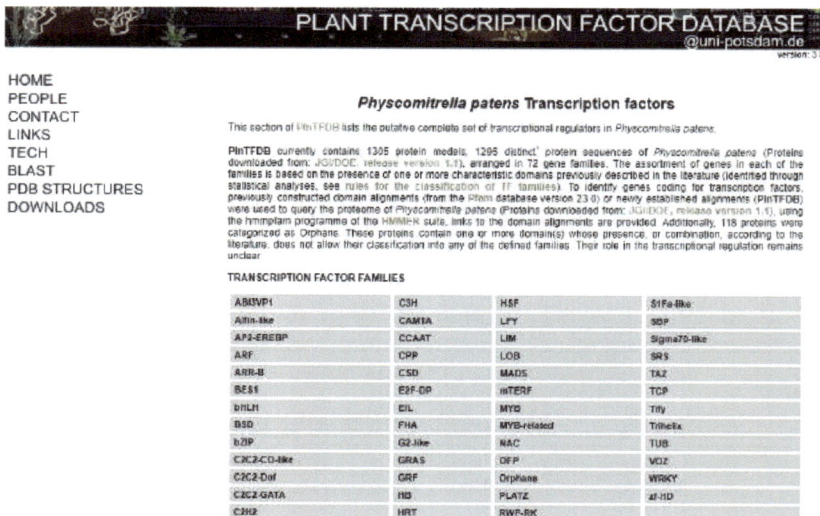

Fig. (2). Plant Transcription Factor Database.

Checklist of Australian Liverworts & Hornworts

Checklist of Australian Liverworts and Hornworts (http://www.anbg.gov.au /abrs/liverwortlist/liverworts_intro.html/; Fig. **3**) contains alphabetically arranged list of accepted taxa under each generic heading of 148 genera of Australian liverworts and hornworts. Synonyms used to Australian specimens are provided with appropriate accepted names. Moreover, distribution using State and Territory is also indicated [65].

Fig. (3). Checklist of Australian Liverworts & Hornworts database resource.

Catalogue of Australian Mosses

Catalogue of Australian Mosses (AusMoss; https://data.rbg. vic.gov.au/cat /mosscatalogue/; Fig. **4**) provides information, including nomenclature, taxonomic, and distribution of Australian mosses and its external territories. Moreover, it also delivers information about continental distribution of Australian species [66].

Australian Mosses Online

Australian Mosses Online (http://www.anbg.gov.au/abrs/Mosses_online/index. html/; Fig. **5**) provides information, including names, synonymy, descriptions, identification keys, ecological data, distribution maps, and illustrations about Australian moss species, genera, and families. The database helps to promote interest, understanding, and significance of the mosses [67].

- Home
- RBG Melbourne
- RBG Cranbourne
- Australian Garden
- About Us
- Gardening Information
- Research and Conservation
- Education
- Support Us

AUSMOSS

Catalogue of Australian Mosses

Search
Provides full wildcard search on genus, epithet, author and more

Browse Alphabetically
Lets you jump straight to a genus by browsing alphabetically

TEXT-ONLY VERSION | PRIVACY POLICY | DISCLAIMER | FOR COMMENTS OR SUGGESTIONS E-MAIL: WEBMASTER@RBG.VIC.GOV.AU

Fig. (4). Catalogue of Australian Mosses database resource.

Fig. (5). Australian Moss Online database resource.

COSMOSS

COSMOSS (http://www.cosmoss.org/; Fig. **6**) is an online database which includes manually curated functional and structural genome annotation for the moss *Physcomitrella patens* [68].

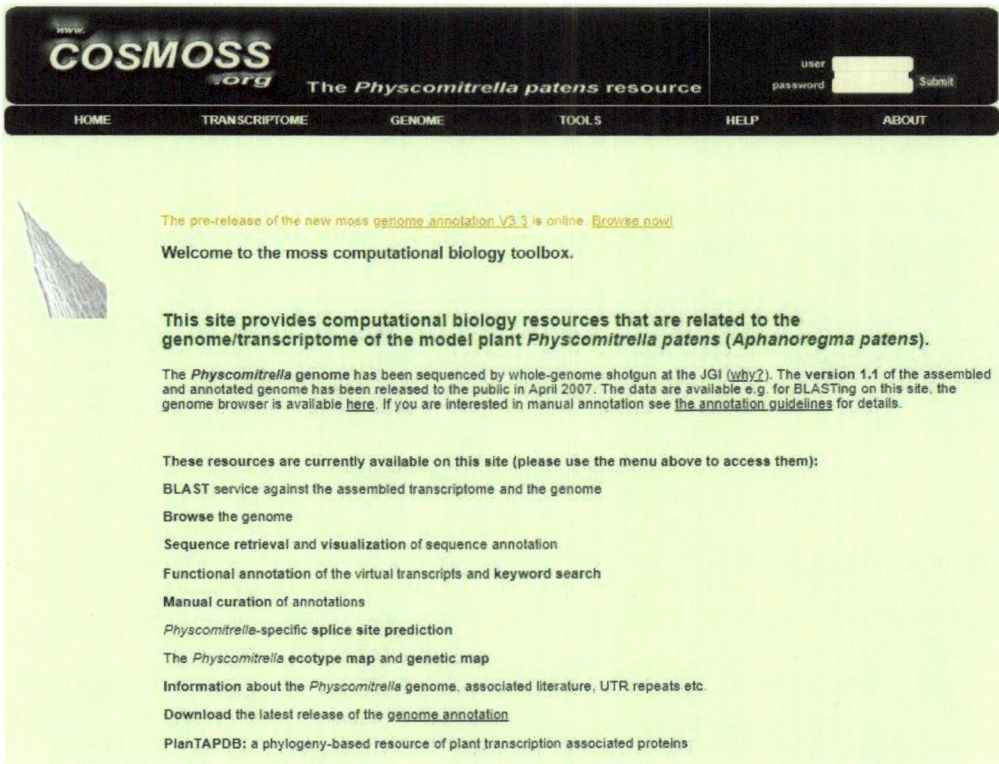

Fig. (6). COSMOSS database.

Plant rDNA Database

The Plant rDNA database (http://www.plantrdnadatabase.com/; Fig. **7**) provides information about numbers and positions of ribosomal DNA (rDNA) loci of bryophytes, angiosperms, gymnosperms, and pteridophytes from publications. Moreover, it also contains information like ploidy level, chromosome number, genome size, life cycle, and telomere composition [69].

BRYOPHYTES

BRYOPHYTES (http://bryophytes.plant.siu.edu/; Fig. **8**) is an online resource dedicated to the field of bryology, which provides information about classification, structural features, natural history, ecology, and evolutionary relationships of bryophyte species.

Fig. (7). Plant rDNA database.

Fig. (8). Bryophyte database resource.

PHYSCObase

The PHYSCObase (http://moss.nibb.ac.jp/; Fig. **9**) is an online resource intended for sharing information about the moss *Physcomitrella patens*. The database provides DNA sequences and expressed sequence tags (EST) of *P. patens*.

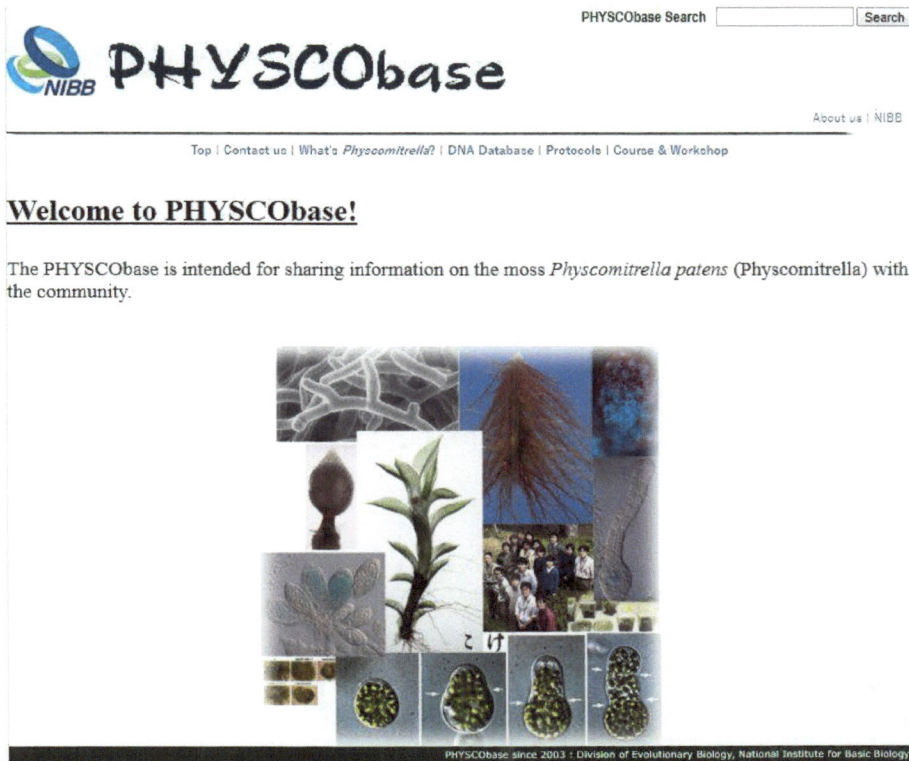

Fig. (9). PHYSCObase database.

Plant DNA C-value Database

The Plant DNA C-value Database (http://www.kew.org/cvalues/; Fig. **10**) contains C-values for 171 bryophyte specie. C-value refers to the DNA amount in the un-replicated gametic nucleus of an organism, irrespective of the ploidy level of the taxon [70].

MitoSatPlant

MitoSatPlant (http://compubio.in/mitosatplant/; Fig. **11**) provides information about simple sequence repeats identified in mitochondrial genome sequences of green plants. Sequences in the database are classified into angiosperms (257), algae (29), bryophytes (8), gymnosperms (1), and pteridophytes (1) [71]. It is evident that organellar genome sequences of bryophytes are poorly represented in public database [72 - 74].

ChloroSSRdb

ChloroSSRdb (www.compubio.in/chlorossrdb/; Fig. **12**) contains information about simple sequence repeats identified in chloroplast genomes of bryophytes. The database also provides comprehensive information of simple sequence repeats mined in chloroplast genomes of algae, angiosperms, gymnosperms, and pteridophytes [75].

Fig. (10). Plant DNA C-value Database.

Various database resources related to bryophyte described in this chapter will play a vital role to extract specific information from specialized resources and will helpful to gain knowledge and information related to bryophytes.

MitoSatPlant: Mitochondrial Microsatellites Database of Viridiplantea

HOME	ABOUT DATABASE		ADVANCED SEARCH		TUTORIAL	STATISTICS		CONTACT

SEQUENCES MINED: 296   ALGAE: 29   BRYOPHYTES: 8   PTERIDOPHYTES: 1   GYMNOSPERMS: 1   ANGIOSPERMS: 257

Total Organisms: 92   ● Perfect ○ Imperfect   Last Updated : 1 Feb 14   Download

ORGANISM	ACCESSION	MONO	DI	TRI	TETRA	PENTA	HEXA	TOTAL
Aegilops speltoides	NC_022666	3	10	15	38	9	5	80
Ajuga reptans	NC_023103	0	2	16	36	6	0	60
Anomodon attenuatus	NC_021931	1	10	4	7	0	0	22

A

Organism: Aegilops speltoides   (NC_022666 , 476091 bp)

Compound : 1   PerfectCompound : 1   Overlapping Compound : 0

Perfect : 80   Density of SSR : ISSR/ 5.81 kb   Average length od SSR : 13.45 bp

Coding : 5   Non-Coding : 76   Coding and Non-Coding : 0

Show Primer   Download

REPEAT	MONO	DI	TRI	TETRA	PENTA	HEXA	TOTAL
PERFECT	3	10	15	38	9	5	80
I.MPERFECT	0	62	210	213	44	29	558

Perfect SSRs

S. No.	MOTIF	LENGTH	START	END	REGION
1.	(T)14	14	209466	209479	Non coding
2.	(T)13	13	238908	238920	Non Coding
3.	(T)13	13	301606	301618	Non coding

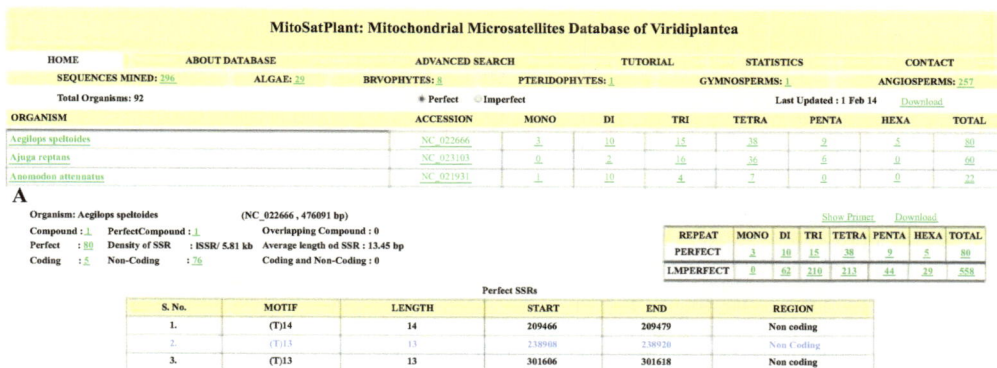

Fig. (11). MitoSatPlant database.

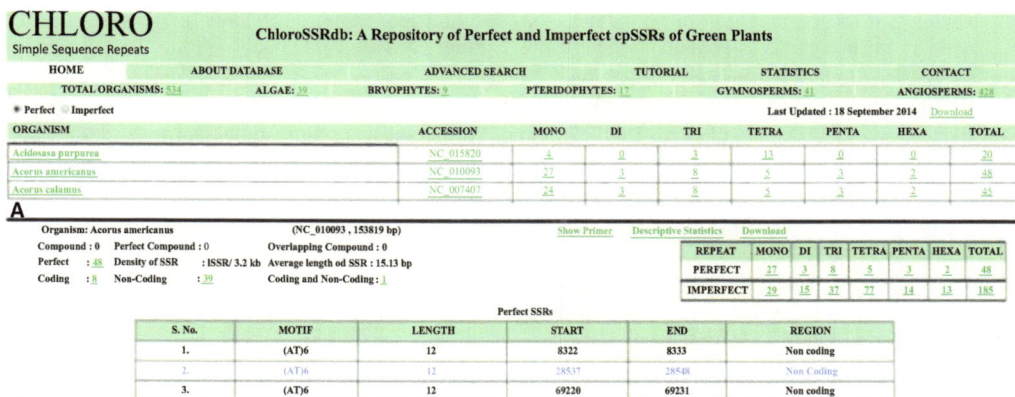

CHLORO
Simple Sequence Repeats

ChloroSSRdb: A Repository of Perfect and Imperfect cpSSRs of Green Plants

HOME	ABOUT DATABASE		ADVANCED SEARCH		TUTORIAL	STATISTICS		CONTACT

TOTAL ORGANISMS: 534   ALGAE: 39   BRYOPHYTES: 9   PTERIDOPHYTES: 17   GYMNOSPERMS: 41   ANGIOSPERMS: 428

● Perfect ○ Imperfect   Last Updated : 18 September 2014   Download

ORGANISM	ACCESSION	MONO	DI	TRI	TETRA	PENTA	HEXA	TOTAL
Acidosasa purpurea	NC_015820	4	0	3	13	0	0	20
Acorus americanus	NC_010093	27	3	8	5	3	2	48
Acorus calamus	NC_007407	24	3	8	5	3	2	45

A

Organism: Acorus americanus   (NC_010093 , 153819 bp)

Compound : 0   Perfect Compound : 0   Overlapping Compound : 0

Perfect : 48   Density of SSR : ISSR/ 3.2 kb   Average length od SSR : 15.13 bp

Coding : 8   Non-Coding : 39   Coding and Non-Coding : 1

Show Primer   Descriptive Statistics   Download

REPEAT	MONO	DI	TRI	TETRA	PENTA	HEXA	TOTAL
PERFECT	27	3	8	5	3	2	48
IMPERFECT	39	15	37	27	14	13	185

Perfect SSRs

S. No.	MOTIF	LENGTH	START	END	REGION
1.	(AT)6	12	8322	8333	Non coding
2.	(AT)6	12	28537	28548	Non Coding
3.	(AT)6	12	69220	69231	Non coding

Fig. (12). ChloroSSRdb database.

CONCLUSION

The different computational resources described here may prove helpful in the study of bryophytes and other organisms. Till date the work related to computational bryology is very limited hence a vast possibilities are there for bryologists.

CONSENT FOR PUBLICATION

Not applicable.

CONFLICT OF INTEREST

The authors confirm that this chapter contents have no conflict of interest.

ACKNOWLEDGEMENTS

SK is thankful to University Grants Commission (UGC), New Delhi, India, for financial support in the form of Rajiv Gandhi National Fellowship.

REFERENCES

[1]   Qiu YL, Li L, Wang B, *et al.* The deepest divergences in land plants inferred from phylogenomic evidence. Proc Natl Acad Sci USA 2006; 103(42): 15511-6.
[http://dx.doi.org/10.1073/pnas.0603335103] [PMID: 17030812]

[2]   Peters K, Gorzolka K, Bruelheide H, Neumann S. Seasonal variation of secondary metabolites in nine different bryophytes. Ecol Evol 2018; 8(17): 9105-17.
[http://dx.doi.org/10.1002/ece3.4361] [PMID: 30271570]

[3]   Wood AJ, Oliver MJ, Cove DJ. Bryophytes as model systems. Bryologist 2000; 103: 128-33.
[http://dx.doi.org/10.1639/0007-2745(2000)103[0128:BAMS]2.0.CO;2]

[4]   Kumar S, Shanker A. Bioinformatics resources for the stress biology of plants In: Vats S, Ed. Biotic and abiotic stress tolerance in plants. Singapore: Springer Nature 2018; pp. 367-86.
[http://dx.doi.org/10.1007/978-981-10-9029-5_14]

[5]   Kumar S, Shanker A. Biological Databases for Medicinal Plant Research In: Kumar N, Ed. Biotechnological Approaches for Medicinal and Aromatic Plants. Singapore: Springer Nature 2018; pp. 655-3.
[http://dx.doi.org/10.1007/978-981-13-0535-1_29]

[6]   Shanker A. Identification of microsatellites in chloroplast genome of *Anthoceros formosae*. Arch Bryol 2013; 191: 1-6.

[7]   Shanker A. Computational mining of microsatellites in the chloroplast genome of *Ptilidium pulcherrimum*, a liverwort. International Journal of Environment 2014; 3: 50-8.
[http://dx.doi.org/10.3126/ije.v3i3.11063]

[8]   Shanker A. Simple sequence repeats mining using computational approach in chloroplast genome of *Marchantia polymorpha*. Arctoa 2014; 23: 145-9.
[http://dx.doi.org/10.15298/arctoa.23.12]

[9]   Shanker A. Detection of simple sequence repeats in the chloroplast genome of *Tetraphis pellucida* Hedw. Plant Sci Today 2016; 3: 207-10.
[http://dx.doi.org/10.14719/pst.2016.3.2.206]

[10]   Zhao CX, Zhu RL, Liu Y. Simple sequence repeats in bryophyte mitochondrial genomes. Mitochondrial DNA A DNA Mapp Seq Anal 2016; 27(1): 191-7.
[http://dx.doi.org/10.3109/19401736.2014.880889] [PMID: 24491104]

[11]   Goruynov DV, Goryunova SV, Kuznetsova OI, *et al.* Complete mitochondrial genome sequence of the "copper moss" *Mielichhoferia elongata* reveals independent nad7 gene functionality loss. PeerJ 2018; 6: e4350.
[http://dx.doi.org/10.7717/peerj.4350]

[12]   Shanker A. Paraphyly of bryophytes inferred using chloroplast sequences. Arch Bryol 2013; 163: 1-5.

[13]   Shanker A. Inference of bryophytes paraphyly using mitochondrial genomes. Arch Bryol 2013; 165: 1-5.

[14]   Shanker A. Combined data from chloroplast and mitochondrial genome sequences showed paraphyly of bryophytes. Arch Bryol 2013; 171: 1-9.

[15]   Schuette S, Piatkowski B, Corley A, Lang D, Geisler M. Predicted protein-protein interactions in the moss Physcomitrella patens: a new bioinformatic resource. BMC Bioinformatics 2015; 16: 89.
[http://dx.doi.org/10.1186/s12859-015-0524-1] [PMID: 25885037]

[16]   Peters K, Gorzolka K, Bruelheide H, Neumann S. Seasonal variation of secondary metabolites in nine different bryophytes. Ecol Evol 2018; 8(17): 9105-17.
[http://dx.doi.org/10.1002/ece3.4361] [PMID: 30271570]

[17]   Sharma V, Munjal A, Shanker A. A text book of Bioinformatics. 2nd Ed., Meerut: Rastogi Publications 2016.

[18]   Benu A, Olivier L. Sequence Alignment in Bioinformatics: Sequences, Structures, Phylogeny. Singapore: Springer Singapore 2018.

[19]   Smith TF, Waterman MS. Identification of common molecular subsequences. J Mol Biol 1981; 147(1): 195-7.
[http://dx.doi.org/10.1016/0022-2836(81)90087-5] [PMID: 7265238]

[20]   Needleman SB, Wunsch CD. A general method applicable to the search for similarities in the amino acid sequence of two proteins. J Mol Biol 1970; 48(3): 443-53.
[http://dx.doi.org/10.1016/0022-2836(70)90057-4] [PMID: 5420325]

[21]   Altschul SF, Gish W, Miller W, Myers EW, Lipman DJ. Basic local alignment search tool. J Mol Biol 1990; 215(3): 403-10.
[http://dx.doi.org/10.1016/S0022-2836(05)80360-2] [PMID: 2231712]

[22]   Altschul SF, Madden TL, Schäffer AA, *et al.* Gapped BLAST and PSI-BLAST: a new generation of protein database search programs. Nucleic Acids Res 1997; 25(17): 3389-402.
[http://dx.doi.org/10.1093/nar/25.17.3389] [PMID: 9254694]

[23]   Pearson WR. Rapid and sensitive sequence comparison with FASTP and FASTA. Methods Enzymol 1990; 183: 63-98.
[http://dx.doi.org/10.1016/0076-6879(90)83007-V] [PMID: 2156132]

[24]   Ashkenazy H, Sela I, Levy Karin E, Landan G, Pupko T. Multiple sequence alignment averaging improves phylogeny reconstruction. Syst Biol 2018.syy036.
[PMID: 29771363]

[25]   Thompson JD, Higgins DG, Gibson TJ. CLUSTAL W: improving the sensitivity of progressive multiple sequence alignment through sequence weighting, position-specific gap penalties and weight matrix choice. Nucleic Acids Res 1994; 22(22): 4673-80.
[http://dx.doi.org/10.1093/nar/22.22.4673] [PMID: 7984417]

[26]   Notredame C, Higgins DG, Heringa J. T-Coffee: A novel method for fast and accurate multiple sequence alignment. J Mol Biol 2000; 302(1): 205-17.
[http://dx.doi.org/10.1006/jmbi.2000.4042] [PMID: 10964570]

[27]   Edgar RC. MUSCLE: multiple sequence alignment with high accuracy and high throughput. Nucleic Acids Res 2004; 32(5): 1792-7.
[http://dx.doi.org/10.1093/nar/gkh340] [PMID: 15034147]

[28]   Katoh K, Kuma K, Toh H, Miyata T. MAFFT version 5: improvement in accuracy of multiple sequence alignment. Nucleic Acids Res 2005; 33(2): 511-8.
[http://dx.doi.org/10.1093/nar/gki198] [PMID: 15661851]

[29]   Morgenstern B. DIALIGN 2: improvement of the segment-to-segment approach to multiple sequence alignment. Bioinformatics 1999; 15(3): 211-8.
[http://dx.doi.org/10.1093/bioinformatics/15.3.211] [PMID: 10222408]

[30]   Beckert S, Steinhauser S, Muhle H, Knoop V. A molecular phylogeny of bryophytes based on nucleotide sequences of the mitochondrial nad5 gene. Plant Syst Evol 1999; 218: 179-92.
[http://dx.doi.org/10.1007/BF01089226]

[31]   Hyvönen J, Koskinen S, Merrill GL, Hedderson TA, Stenroos S. Phylogeny of the Polytrichales (Bryophyta) based on simultaneous analysis of molecular and morphological data. Mol Phylogenet Evol 2004; 31(3): 915-28.

[http://dx.doi.org/10.1016/j.ympev.2003.11.003] [PMID: 15120390]

[32]   Shanker A, Sharma V, Daniell H. A novel index to identify unbiased conservation between proteomes. Int J Integr Biol 2009; 7: 32.

[33]   Pitman MR, Menz RI. Methods for protein homology modelling. Applied Mycology and Biotechnology 2006; 6: 37-59.
[http://dx.doi.org/10.1016/S1874-5334(06)80005-5]

[34]   Schwede T, Kopp J, Guex N, Peitsch MC. SWISS-MODEL: An automated protein homology-modeling server. Nucleic Acids Res 2003; 31(13): 3381-5.
[http://dx.doi.org/10.1093/nar/gkg520] [PMID: 12824332]

[35]   Guex N, Peitsch MC. SWISS-MODEL and the Swiss-PdbViewer: an environment for comparative protein modeling. Electrophoresis 1997; 18(15): 2714-23.
[http://dx.doi.org/10.1002/elps.1150181505] [PMID: 9504803]

[36]   Webb B, Sali A. Protein structure modeling with MODELLER. Methods Mol Biol 2014; 1137: 1-15.
[http://dx.doi.org/10.1007/978-1-4939-0366-5_1] [PMID: 24573470]

[37]   Lambert C, Léonard N, De Bolle X, Depiereux E. ESy Pred3D: Prediction of proteins 3D structures. Bioinformatics 2002; 18(9): 1250-6.
[http://dx.doi.org/10.1093/bioinformatics/18.9.1250] [PMID: 12217917]

[38]   Kim DE, Chivian D, Baker D. Protein structure prediction and analysis using the Robetta server. Nucleic Acids Res 2004; 32(Web Server issue)W526-31.
[http://dx.doi.org/10.1093/nar/gkh468] [PMID: 15215442]

[39]   Jayaram B, Bhushan K, Shenoy SR, *et al.* Bhageerath: an energy based web enabled computer software suite for limiting the search space of tertiary structures of small globular proteins. Nucleic Acids Res 2006; 34(21): 6195-204.
[http://dx.doi.org/10.1093/nar/gkl789] [PMID: 17090600]

[40]   Wu S, Zhang Y. LOMETS: a local meta-threading-server for protein structure prediction. Nucleic Acids Res 2007; 35(10): 3375-82.
[http://dx.doi.org/10.1093/nar/gkm251] [PMID: 17478507]

[41]   Nielsen M, Lundegaard C, Lund O, Petersen TN. CPH models-3.0-remote homology modeling using structure-guided sequence profiles. Nucleic Acids Res 2010; 38(Web Server issue)W576-81.
[http://dx.doi.org/10.1093/nar/gkq535] [PMID: 20542909]

[42]   Blaszczyk M, Jamroz M, Kmiecik S, Kolinski A. CABS-fold: Server for the de novo and consensus-based prediction of protein structure. Nucleic Acids Res 2013; 41(Web Server issue)W406-11.
[http://dx.doi.org/10.1093/nar/gkt462] [PMID: 23748950]

[43]   Källberg M, Margaryan G, Wang S, Ma J, Xu J. Raptor X server: a resource for template-based protein structure modeling. Protein Structure Prediction. New York Humana Press 2014; pp. 17-27.
[http://dx.doi.org/10.1007/978-1-4939-0366-5_2]

[44]   Shen Y, Maupetit J, Derreumaux P, Tufféry P. Improved PEP-FOLD approach for peptide and miniprotein structure prediction. J Chem Theory Comput 2014; 10(10): 4745-58.
[http://dx.doi.org/10.1021/ct500592m] [PMID: 26588162]

[45]   Yang J, Yan R, Roy A, Xu D, Poisson J, Zhang Y. The I-TASSER Suite: protein structure and function prediction. Nat Methods 2015; 12(1): 7-8.
[http://dx.doi.org/10.1038/nmeth.3213] [PMID: 25549265]

[46]   Singh A, Kaushik R, Mishra A, Shanker A, Jayaram B. ProTSAV: A protein tertiary structure analysis and validation server. Biochim Biophys Acta 2016; 1864(1): 11-9.
[http://dx.doi.org/10.1016/j.bbapap.2015.10.004] [PMID: 26478257]

[47]   Morris GM, Huey R, Lindstrom W, *et al.* AutoDock4 and AutoDockTools4: Automated docking with selective receptor flexibility. J Comput Chem 2009; 30(16): 2785-91.

[http://dx.doi.org/10.1002/jcc.21256] [PMID: 19399780]

[48]   Trott O, Olson AJ. AutoDock Vina: improving the speed and accuracy of docking with a new scoring function, efficient optimization, and multithreading. J Comput Chem 2010; 31(2): 455-61.
[PMID: 19499576]

[49]   Friesner RA, Banks JL, Murphy RB, *et al.* Glide: a new approach for rapid, accurate docking and scoring. 1. Method and assessment of docking accuracy. J Med Chem 2004; 47(7): 1739-49.
[http://dx.doi.org/10.1021/jm0306430] [PMID: 15027865]

[50]   Yang JM, Chen CC. GEMDOCK: a generic evolutionary method for molecular docking. Proteins 2004; 55(2): 288-304.
[http://dx.doi.org/10.1002/prot.20035] [PMID: 15048822]

[51]   Jayaram B, Singh T, Mukherjee G, Mathur A, Shekhar S, Shekhar V. Sanjeevini: a freely accessible web-server for target directed lead molecule discovery. BMC Bioinformatics 2012; 13 (Suppl. 17): S7.
[http://dx.doi.org/10.1186/1471-2105-13-S17-S7] [PMID: 23282245]

[52]   Pearlman DA, Case DA, Caldwell JW, *et al.* AMBER, a package of computer programs for applying molecular mechanics, normal mode analysis, molecular dynamics and free energy calculations to simulate the structural and energetic properties of molecules. Comput Phys Commun 1995; 91: 1-41.
[http://dx.doi.org/10.1016/0010-4655(95)00041-D]

[53]   Brooks BR, Bruccoleri RE, Olafson BD, States DJ, Swaminathan SA, Karplus M. CHARMM: a program for macromolecular energy, minimization, and dynamics calculations. J Comput Chem 1983; 4: 187-217.
[http://dx.doi.org/10.1002/jcc.540040211]

[54]   Hess B, Kutzner C, van der Spoel D, Lindahl E. GROMACS 4: algorithms for highly efficient, load-balanced, and scalable molecular simulation. J Chem Theory Comput 2008; 4(3): 435-47.
[http://dx.doi.org/10.1021/ct700301q] [PMID: 26620784]

[55]   Attwood TK, Gisel A, Eriksson NE, Bongcam-Rudloff E. Concepts, historical milestones and the central place of bioinformatics in modern biology: a European perspective. China: Bioinformatics-Trends and Methodologies. InTech 2011.

[56]   Benson DA, Cavanaugh M, Clark K, *et al.* GenBank. Nucleic Acids Res 2013; 41(Database issue): D36-42.
[PMID: 23193287]

[57]   Barker WC, Garavelli JS, Haft DH, *et al.* The PIR-international protein sequence database. Nucleic Acids Res 1998; 26(1): 27-32.
[http://dx.doi.org/10.1093/nar/26.1.27] [PMID: 9399794]

[58]   Wheeler DL, Barrett T, Benson DA, *et al.* Database resources of the national center for biotechnology information. Nucleic Acids Res 2008; 36(Database issue): D13-21.
[PMID: 18045790]

[59]   Boutet E, Lieberherr D, Tognolli M, Schneider M, Bairoch A. Uniprotkb/swiss-prot. Plant Bioinformatics 2007; pp. 89-112.

[60]   Berman HM, Westbrook J, Feng Z, *et al.* The protein data bank. Nucleic Acids Res 2000; 28(1): 235-42.
[http://dx.doi.org/10.1093/nar/28.1.235] [PMID: 10592235]

[61]   Chen J, Anderson JB, DeWeese-Scott C, *et al.* MMDB: Entrez's 3D-structure database. Nucleic Acids Res 2003; 31(1): 474-7.
[http://dx.doi.org/10.1093/nar/gkg086] [PMID: 12520055]

[62]   Kanehisa M, Goto S. KEGG: kyoto encyclopedia of genes and genomes. Nucleic Acids Res 2000; 28(1): 27-30.
[http://dx.doi.org/10.1093/nar/28.1.27] [PMID: 10592173]

[63]   Bolser D, Staines DM, Pritchard E, Kersey P. Ensembl plants: integrating tools for visualizing, mining, and analyzing plant genomics data. Plant Bioinformatics 2016; pp. 115-40.

[64]   Pérez-Rodríguez P, Riaño-Pachón DM, Corrêa LG, Rensing SA, Kersten B, Mueller-Roeber B. PlnTFDB: updated content and new features of the plant transcription factor database. Nucleic Acids Res 2010; 38(Database issue): D822-7.
[http://dx.doi.org/10.1093/nar/gkp805] [PMID: 19858103]

[65]   McCarthy PM. Checklist of Australian Liverworts and Hornworts. 2006.

[66]   Klagenza N, Streimann H, Curnow J. AUSMOSS Catalogue of Australian mosses. Melbourne: Royal Botanic Gardens 2009.

[67]   McCarthy PM. Australian Mosses Online. 2013.

[68]   Zimmer AD, Lang D, Buchta K, *et al.* Reannotation and extended community resources for the genome of the non-seed plant Physcomitrella patens provide insights into the evolution of plant gene structures and functions. BMC Genomics 2013; 14: 498.
[http://dx.doi.org/10.1186/1471-2164-14-498] [PMID: 23879659]

[69]   Garcia S, Gálvez F, Gras A, Kovařík A, Garnatje T. Plant rDNA database: update and new features. Database (Oxford) 2014; 2014 bau063.
[http://dx.doi.org/10.1093/database/bau063] [PMID: 24980131]

[70]   Greilhuber J, Obermayer R, Leitch IJ, Bennett MD. Bryophyte DNA C-values database (release 30). 2010.

[71]   Kumar M, Kapil A, Shanker A. MitoSatPlant: mitochondrial microsatellites database of viridiplantae. Mitochondrion 2014; 19(Pt B): 334-7.
[http://dx.doi.org/10.1016/j.mito.2014.02.002] [PMID: 24561221]

[72]   Shanker A. Chloroplast genomes of bryophytes: a review. Arch Bryol 2012; p. 143.

[73]   Shanker A. Sequenced mitochondrial genomes of bryophytes. Arch Bryol 2012; p. 146.

[74]   Shanker A. An update on sequenced chloroplast genomes of Bryophytes. Plant Sci Today 2015; 2: 172-4.
[http://dx.doi.org/10.14719/pst.2015.2.4.143]

[75]   Kapil A, Rai PK, Shanker A. ChloroSSRdb: a repository of perfect and imperfect chloroplastic simple sequence repeats (cpSSRs) of green plants. Database (Oxford) 2014; 2014 bau107.
[http://dx.doi.org/10.1093/database/bau107] [PMID: 25380781]

CHAPTER 4

The Utility of Molecular Sequence Data in Phylogenetic Analysis of Bryoflora

Saumya Pandey and **Afroz Alam**[*]

Bryotechnology Laboratory, Department of Bioscience and Biotechnology, Banasthali Vidyapith, Rajasthan- 304022, India

Abstract: The molecular sequence data have been utilized immensely to resolve the phylogeny of bryophytes, primarily the monophyletic relationship of bryophytes and their position in the evolution of land plants. However, the study also described division bryophyta as a paraphyletic group with only one of its classes related to the vascular plants and another one sister to all other land plants. The debatable position of earliest diverging lineages of the mosses such as *Takakia* and *Sphagnum* spp., and other genera can also be resolved more accurately with increasing knowledge of fast DNA sequencing techniques and bioinformatics tools. Different nuclear-encoded genes (18S, 26S) or chloroplast genes (*trn*F-*trn*L, *psb*A, *rbc*L, *rps*4) and mitochondrial encoded genes (*Cob* intron, *nad*2, *nad*5) emerge as the suitable marker to understand the deep level of the molecular phylogeny of bryophytes. The selection of suitable markers, evolutionary model and phylogenetic tree evaluation methods are necessary for the better understanding of molecular phylogeny.

Keywords: Bryophytes, DNA sequencing, Molecular phylogeny, Phylogenetic tree.

INTRODUCTION

The genome of organisms holds very useful information about taxonomy, phylogenies, biogeography, and population dynamics. So the characterization of the organism at the molecular level may provide a reliable and replicable tool leading to the identification of an organism, like barcoding, and/or to the description of the evolutionary relationships among individual/taxa. The advent of DNA sequencing has offered the possibility of independent phylogenies, unbiased by human interpretation of morphology. Sequencing method compares homologous loci so it is more reliable than markers based on the banding pattern (Fig. **1**). *Marchantia polymorpha 5S* gene sequence was the first DNA sequence obtained for bryophytes [1]. However, the first studies on bryophyte evolution

---

[*] **Corresponding author Afroz Alam**, Bryotechnology Laboratory, Department of Bioscience and Biotechnology, Banasthali Vidyapith, Rajasthan, India; Tel: +91-9785453594; E-mail: afrozalamsafvi@gmail.com

Afroz Alam (Ed.)

were carried out using genes encoding ribosomal RNAs [2, 3]. The nuclear ribosomal DNA is widely used to reconstruct phylogenetic relationships [4, 5], within families [6], genera [7, 8] or species [9]. Particularly, ITS sequences were used as markers in many studies in mosses [10].

The chloroplast (cp) *trn*T-F region and especially the *trn* (UAA) intron are the most widely targeted loci, not only in bryophytes but also in other plants [11]. The locus has been complemented by the *rps*4 gene in many studies aiming at resolving relationships within classes and families [12, 13], genera [7] and even among populations [9]. The *atp*B-*rbc*L intergenic spacer has also been frequently used in moss phylogeography [14]. The *rsp*4 and *trn*F have been also used to clarify systematic affinities [15]. The mitochondrial gene, *nad*5, has been used for reconstructing bryophyte phylogenies [16], and so have other mitochondrial DNA regions [17 - 19].

The Classical Method of Phylogenetic Analysis of Bryoflora and its Drawback

Traditionally, the identification and classifications of bryophytes were solely relying on gametophytic and sporophytic characters [6]. However, the study based on the sporophytic characters such as the position of perichaetium, the position of sporophytes (acrocarpous or pleurocarpous), peristome development and structure, etc. were considered more accurate and preferred over gametophytic characters. But nowadays molecular phylogenetics is gaining popularity and became the most widely used method for bryophyte systematic [20].

The diversity determination can be based on morphological, biochemical, and molecular types of information [21]. These morphological characters are usually studied by using a lens, Camera Lucida, light microscope, and scanning light microscope. However, morpho-anatomical characters based study are (a) time-consuming, (b) uncertain and, (c) the characters may vary in response to ecological and environmental condition, thus molecular markers have advantages over other kinds as they show genetic differences on a more detailed level without interferences from environmental factors and involve techniques that provide fast results detailing genetic diversity [22].

The Essential Characteristic of Molecular Marker Gene

The essential features of the molecular marker gene are listed below:

Single-copy Genes: Single-copy genes have been identified as a more useful marker than multiple-copy gene in studying the phylogeny of unresolved lineages. The single-copy genes are unique and highly conserved sequence across the species and thus are easily amplified and sequenced. The nuclear genes are preferred choice to study the speciation and phylogenetic relationship as they have the bi-parental inheritance in contrast to organellar gene that is uni-parentally inherited [23].

Easy Sequence Alignment: The alignments of marker gene sequences are important for the phylogenetic analysis. The length of the same gene can be variable among different member of taxa because of insertions and deletions and thus sequence alignment may become difficult. Hence, for accurate phylogeny, the regions with ambiguous alignment can be particularly excluded from study or secondary structure information of gene can be applied [24].

Optimum Substitution Rate: The gene with optimum substitution rate should be preferred as the multiple substitutions may lead to the state of saturation. The multiple substitutions can generate a non-phylogenetic signal causing misinterpretation of the relationship among the groups [25].

Primer Specificity: The primer should specifically amplify the marker gene. The primers that are too universal must be avoided as they may also produce non-specific amplicons due to the presence of contaminants or symbionts [26].

Molecular Markers Genes and its Utility in the Phylogeny of Bryoflora

The first basic requirement for the sequencing-based analysis of phylogenetic history is a selection of the slowly evolving gene or amino acid sequences. All the genes cannot be appropriately used as phylogenetic markers. Similarly, not all markers can be successfully utilized for the analysis of the given group of organisms. The utility of genes to determine the relationship within particular groups can be studied using empirical saturation plots. This plot shows the rate of variation in informative character during the geological time as a clade undergone phylogenetic divergence [27].

Corresponding to the particular divergences; specific genes must be selected [28]. Thus, for the analysis of the large groups, whose divergence span the longer extent of time, many gene sequences may be required to resolve different regions of the tree, the structural information of genes or different sets of sites within genes may provide a better understanding of phylogeny [29].

Nuclear Ribosomal Genes

Several reports have been published which mainly focuses on the molecular organization and efficiency of nuclear rDNA especially *ITS* markers in the phylogenetic study of bryophytes [30, 31]. In addition, there is a gradual increase in the number of other nuclear markers for phylogenetic analysis of bryophytes such as *adk* [32], *gpd* [33], *LFY* [34, 35], *phy2* [36], and *rpb2* [37]. However, none of these nuclear markers has gained popularity as barcoding markers because sequencing and amplification require taxon-specific primers and cloning after initial PCR. Most of these nuclear genes were sequenced from peat moss *Sphagnum* [38, 39]. Some of the selected nuclear ribosomal genes used in the phylogenetic study of bryophytes are explained below in some detail.

18S rRNA Gene: The most widely used markers to study phylogenetic relationships within land plants in the 19th century. However, it has not been used to study the relationship between and within genera because it shows limited sequence variation below the family level. *18S* rRNA gene has been used to infer phylogeny in moss lineages [40 - 42] and liverwort [43, 44]. Also, *18S* rRNA based phylogenetic analyses showed ambiguity regarding monophyly or paraphyly [45] of liverworts. This uncertainty is mainly due to incomplete taxon sampling and the use of *18S* gene alone.

26S: The first attempt for the sequencing of this locus was done by Waters *et al.* [3]. Thereafter, *26S* gene gain popularity in bryophyte systematic [46]. Also, in contrast to *18S* gene, *26S* gene sequence has been successfully used at family and genus level to resolve the phylogeny among various groups of both mosses [47, 12, 48] and liverworts [44].

ITS Region: The nuclear ribosomal internal transcribed spacers (*ITS*1-5.8S-*ITS*2) are extensively exploited nuclear markers for the phylogenetic inferences in bryophytes [49]. Recently, Biersma *et al.* [50] used both *ITS1* and *ITS2* to study the diversity, richness and relative age divergences within most specious plant genus *Schistidium* in Antartica. The result obtained partially support the intergeneric classification of Antarctic *Schistidium* species as given by Ochyra *et al.* [51]. Similarly, ITS region along with chloroplast markers has been used to resolve the taxonomic position of genus *Pohilia*. The phylogenetic analysis revealed that genus *Pohilia* non-monophyletic and closer to the genera in Mielichhoferiaceae and Mniaceae [52].

rpb2 Gene: This gene codes for RNA polymerase II subunit 2 (*rbp2*). Fuselier *et al.* [37] used partial *rpb2* gene region to study the phylogenetic and phylogeographic relationship among three *Metzgeria* species and showed the efficiency of this marker in resolving species-level relationship in liverworts. Recently, *rpb* gene along with *LFY* gene has been used to study the hybrid formation and genetic diversity with genus *Diphasiastrum* [35].

Mitochondrial Genes

MtDNA has been extensively used in phylogenetic analysis of unresolved lineages of bryoflora because of (a) high conservation, (b) slow sequence evolution comparable to nuclear and chloroplast gene, (c) significant RNA editing and trans-splicing than in chloroplast, (d) stable localization of introns, (e) maternal inheritance and, (f) higher mutation rate than nuclear DNA [53]. Till now 51 mitochondrial complete genomes are sequenced for bryophyte mostly belonging to just four families Funariaceae [54], Grimmiaceae [55, 56], Orthotrichaceae [57 - 59] and Sphagnaceae [60].

nad Gene: The *nad5-nad4-nad2* gene cluster is conserved among mosses, hornwort, and liverwort [61]. It contains two spacers and a variable number of introns. Further, based on the secondary structure and splicing mechanism, the introns are of two types group 1 and group 2. The *nad5-nad4* IGS can be used as an ideal marker for the phylogenetic analysis of bryoflora as the length of conserved sequences of *nad5-nad4* IGS is nearly 600 bp in mosses, 1,000–1,300 bp in liverworts, and above 3,000 bp in hornworts whereas for *nad4-nad2* IGS the size of the conserved sequence is 26 bp in all three bryophyte lineages [19]. Beckert *et al.* [16] reported the relationship among liverwort and mosses based on group 1 (G1) intron sequence conserved in *nad5* gene. Similarly, the *nad2* genes which contain only group 2 (G2) introns and are positionally conserved over large evolutionary distances have been used to study the phylogenetic relationship among broad taxa of mosses [62]. Both these genes (*nad2* and *nad5*) code for protein subunits of the NADH dehydrogenase (complex I of the mitochondrial respiratory chain). However, among all mitochondrial markers group 1 intron has gained popularity among researcher to study the relationship among bryophytes lineages. The *nad5* G1 intron has been used to provide a better understanding of higher-level phylogenetic relationships of mosses [63, 18] and liverworts [47, 64]. Also, the reconstruction of phylogeny based on mitochondrial genes *nad5* or *nad2* showed slightly increased support values as compared to *rbcL* or *rps4* [19].

cob and cox Gene: The *cox* and *cob* genes are universally located in the mitochondrial genome of all land plants and codes for the protein subunit of cytochrome bc1 complex and cytochrome c oxidase, respectively [65]. The *cox3* gene is the first sequenced mitochondrial markers and primarily utilized to infer relationships among land plants [66, 67]. Also, *cox3* gene has been successfully utilized to study organellar inheritance in liverworts [68]. In addition, two group 1 introns, *cox1* and *cobi420* have been successfully identified within complete mitochondrial genome sequence of moss *Physcomitrella patens* [69]. These genes have been exploited by the various researcher for a better understanding of bryophyte phylogenetics [70].

Chloroplast Genes

In bryophytes, most of the phylogenetic studies have been based on the chloroplast genes as they are highly conserved and show very little rearrangements. Till now, the complete chloroplast genome is sequenced for 14 bryophyte species [71]. Currently, a large number of chloroplast genes has been used to study relationship among several lineages of both liverwort and mosses such as *rps4* [72, 73], *rbcL* [74], *atpB-rbcL* spacer [75], *trnG$_{UCC}$* G2 intron [37], *rpl16*, *trnK/matK*, *rpoA*, *accD-psaI*, *atpB* [67], *trnM-V* [76], *trnL-trn*F [77], *psaA*, *psaB*, *psbD* [78], *16S* [6] and *23S* [67] gene and many more. Some commonly used plastid markers are explained in details.

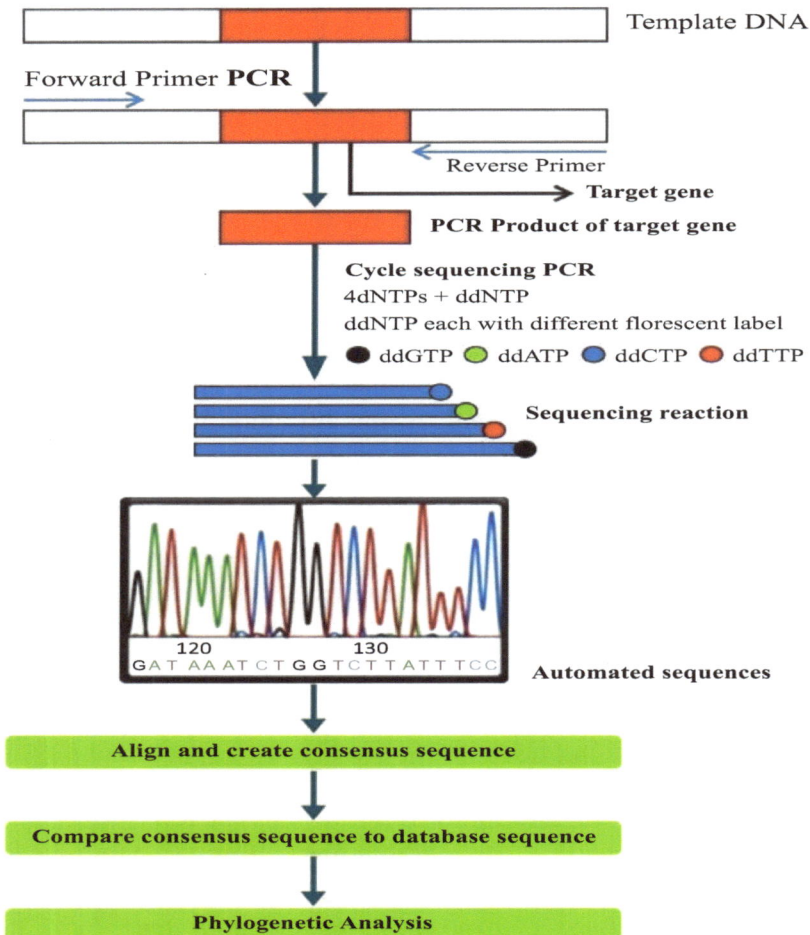

Fig. (1). Diagrammatic representation of sequencing based phylogenetic analysis.

16S and 23S rRNA Genes: Till now only a few reports are available which have utilized *16S* and *23S* rRNA sequence for phylogenetic analysis of bryophytes [2] and land plants [67, 78]. Similarly, Pandey *et al.* [6] used partial *16S* rRNA gene to infer controversial phylogenies of the moss lineages.

rps 4 Gene: *rps4* is one of the most widely used plastid markers to study the phylogenetic relationship in bryophytes. The study suggests that the phylogenetic tree obtained by *rps4* gene is comparatively more parsimony informative than *rbcL* gene [79]. The *rps4* gene has been widely used to resolve relationship among mosses [50, 80] as well as liverwort [81, 82] at ordinal, family and genus level [83, 84]. Biersma *et al.* [50] used multilocus approach (nuclear ITS, *trnL-F* and *rps4*) to investigate the genetic variation between bank-forming moss *Chorisodontium aciphyllum*. Similarly, Yi *et al.* [72] used *rps4* and *ITS2* gene for identification of three morphologically similar species of genus *Plagiomnium*.

rbcL: The *rbcL* gene was one of the first sequenced plant genes and is widely used for inferring relationship among land plants [85]. However, due to low sequence variation *rbcL* is only used at the family level and unsuitable for the species and population level analysis of relationship within bryophytes [18]. Banting *et al.* [74] used only *rbcL* gene for phylogenetic analysis of liverwort and reported well-supported sister relationship among various groups of liverworts.

atpB-rbcL Spacer: It is one of the most commonly used plastid spacer. The sequence length of the *atpB-rbcL* spacer is comparatively shorter in bryophytes (approx. 300-700 nt) than in seed plants (approx. 800–1,000 nt) [86]. However, in contrast to other commonly used plastid spacers (*trnL-F*, *psbA-trnH* and *rps4-trnS*), *atpB-rbcL* longer sequence length and contain more informative sites for better resolution of the clade. This gene has been widely used at a different taxonomic level ranging from class level to intraspecific level [86]. However, mostly used to study the phylogenetic relationship at family and the genus level in both liverwort and mosses [87].

trnK/matK Region: First attempt to study bryophytes systematic using *trnK/matK* region was done to resolve the relationships in genus *Asterella* [88]. The *trnK/matK* region is usually efficient to resolve the phylogeny at genus level or above. Justin *et al.* [89] used nuclear and plastid marker (*matK* and *rpl16* intron) to resolve the relationship within pleurocarpous moss genus *Plagiothecium*. The plastid markers provide better resolutions of the clade as compared to that obtained from nuclear data.

CONCLUSION

The molecular sequence data (nuclear, mitochondrial and chloroplast genes) has been successfully utilized to study the phylogenetic relationship among various lineages among bryophytes. However, complete genome sequences of only few bryophytes species available. Thus, the increase in whole genome sequencing of bryophyte species will facilitate the detection of the more potential gene for the phylogenetic studies which can be further employed to study unresolved taxa.

CONSENT FOR PUBLICATION

Not applicable.

CONFLICT OF INTEREST

The authors confirm that this chapter contents have no conflict of interest.

ACKNOWLEDGEMENT

The authors are grateful to Prof. Aditya Shastri, Vice Chancellor, Banasthali Vidyapith, Rajasthan, for his encouragement and support.

REFERENCES

[1]   Hori H, Lim BL, Osawa S. Evolution of green plants as deduced from 5S rRNA sequences. Proc Natl Acad Sci USA 1985; 82(3): 820-3.
[http://dx.doi.org/10.1073/pnas.82.3.820] [PMID: 16593540]

[2]   Mishler BD, Thrall PH, Hopple JS, De-Luna E, Vilalys R. A molecular approach to the phylogeny of bryophytes: Cladistic analysis of chloroplast encoded 16S and 23S ribosomal RNA genes. Bryologist 1992; 95: 172-80.
[http://dx.doi.org/10.2307/3243432]

[3]   Waters DA, Buchheim MA, Dewey RA, Chapman RL. Preliminary inferences of the phylogeny of bryophytes from nuclear-encoded ribosomal RNA Sequences. Am J Bot 1992; 79: 459-66.
[http://dx.doi.org/10.1002/j.1537-2197.1992.tb14575.x]

[4]   Olsson S, Buchbender V, Enroth J, Hedenäs L, Huttunen S, Quandt D. Phylogenetic analyses reveal high levels of polyphyly among pleurocarpous lineages as well as novel clades. Bryologist 2009; 112: 447-66.
[http://dx.doi.org/10.1639/0007-2745-112.3.447]

[5]   Merget B, Wolf M. A molecular phylogeny of Hypnales (Bryophyta) inferred from ITS2 sequence-structure data. BMC Res Notes 2010; 3: 320-7.
[http://dx.doi.org/10.1186/1756-0500-3-320] [PMID: 21108782]

[6]   Pandey S, Sharma V, Alam A. Phylogeny based on 16S rRNA sequence and morphology of selected mosses of Mount Abu, Rajasthan (India). Meta Gene 2018; 16: 218-25.
[http://dx.doi.org/10.1016/j.mgene.2018.03.006]

[7]   Shaw AJ, Cox CJ, Buck WR, *et al.* Newly resolved relationships in an early land plant lineage: Bryophyta class Sphagnopsida (peat mosses). Am J Bot 2010; 97(9): 1511-31.
[http://dx.doi.org/10.3732/ajb.1000055] [PMID: 21616905]

[8]   Carter BE. Species delimitation and cryptic diversity in the moss genus *Scleropodium* (Brachytheciaceae). Mol Phylogenet Evol 2012; 63(3): 891-903.
[http://dx.doi.org/10.1016/j.ympev.2012.03.002] [PMID: 22421213]

[9]   Draper I, González-Mancebo JM, Werner O, Patiño J, Ros RM. Phylogeographic relationships between the mosses *Exsertotheca intermedia* from Macaronesian Islands and *Neckera baetica* from southern glacial refugia of the Iberian Peninsula. Ann Bot Fenn 2011; 48: 133-41.
[http://dx.doi.org/10.5735/085.048.0205]

[10]   Terracciano S, Giordano S, Spagnuolo V. A further tessera in the two centuries old debate on the *Hypnum cupressiforme* complex (Hypnaceae, Bryopsida). Plant Syst Evol 2012; 298: 229-38.
[http://dx.doi.org/10.1007/s00606-011-0540-1]

[11]   Quandt D, Stech M. Molecular evolution of the *trn*L(UAA) intron in bryophytes. Mol Phylogenet Evol 2005; 36(3): 429-43.
[http://dx.doi.org/10.1016/j.ympev.2005.03.014] [PMID: 16005648]

[12]   Pedersen N, Holyoak DT, Newton AE. Systematics and morphological evolution within the moss family Bryaceae: A comparison between parsimony and Bayesian methods for reconstruction of ancestral character states. Mol Phylogenet Evol 2007; 43(3): 891-907.
[http://dx.doi.org/10.1016/j.ympev.2006.10.018] [PMID: 17161629]

[13]   Bell NE, Hyvönen J. Phylogeny of the moss class *Polytrichopsida* (BRYOPHYTA): Generic-level structure and incongruent gene trees. Mol Phylogenet Evol 2010; 55(2): 381-98.
[http://dx.doi.org/10.1016/j.ympev.2010.02.004] [PMID: 20152915]

[14]   Grundmann M, Schneiderb H, Russella SJ, Vogela JC. Phylogenetic relationships of the moss genus *Pleurochaete* Lindb. (Bryales: Pottiaceae) based on chloroplast and nuclear genomic markers. Org Divers Evol 2006; 6: 33-45.
[http://dx.doi.org/10.1016/j.ode.2005.04.005]

[15]   Hernández-Maqueda R, Quandt D, Werner O, Muñoz J. Phylogeny and classification of the Grimmiaceae/Ptychomitriaceae complex (Bryophyta) inferred from cpDNA. Mol Phylogenet Evol 2008; 46(3): 863-77.
[http://dx.doi.org/10.1016/j.ympev.2007.12.017] [PMID: 18262799]

[16]   Beckert S, Steinhauser S, Muhle H, Knoop V. A Molecular Phylogeny of bryophytes based on nucleotide sequence of the mitochondria nad5 gene. Plant Syst Evol 1999; 218: 179-92.
[http://dx.doi.org/10.1007/BF01089226]

[17]   Cox CJ, Goffinet B, Wickett NJ, Boles SB, Shaw AJ. Moss diversity: A molecular phylogenetic analysis of genera. Phytotaxa 2010; 9: 175-95.
[http://dx.doi.org/10.11646/phytotaxa.9.1.10]

[18]   Wahrmund U, Quandt D, Knoop V. The phylogeny of mosses - addressing open issues with a new mitochondrial locus: Group I intron cobi420. Mol Phylogenet Evol 2010; 54(2): 417-26.
[http://dx.doi.org/10.1016/j.ympev.2009.09.031] [PMID: 19853052]

[19]   Wahrmund U, Rein T, Muller KF, Groth-Malonek M, Knoop V. Fifty mosses on five trees: Comparing phylogenetic information in three types of non-coding mitochondrial DNA and two chloroplast loci. Plant Syst Evol 2009; 282: 241-55.
[http://dx.doi.org/10.1007/s00606-008-0043-x]

[20]   Goffinet B, Buck WR, Shaw AJ, Eds. Morphology, anatomy, and classification of the Bryophyta. USA: New York: Cambridge University Press 2009; pp. 55-138.

[21]   Gonçalves LS, Rodrigues R, do Amaral Júnior AT, Karasawa M, Sudré CP. Heirloom tomato gene bank: assessing genetic divergence based on morphological, agronomic and molecular data using a Ward-modified location model. Genet Mol Res 2009; 8(1): 364-74.
[http://dx.doi.org/10.4238/vol8-1gmr549] [PMID: 19440972]

[22]   Saker MM, Youssef SS, Abdallah NA, Bashandy HS. Genetic analysis of some Egyptian rice genotypes using RAPD, SSR and AFLP. Afr J Biotechnol 2005; 4: 882-90.

[23]   Li Z, De La Torre AR, Sterck L, *et al.* Single-copy gene as molecular markers for phylogenomic studies in seed plants. Genome Biol Evol 2017; 9(5): 1130-47.

[http://dx.doi.org/10.1093/gbe/evx070] [PMID: 28460034]

[24]   Kjer KM. Use of rRNA secondary structure in phylogenetic studies to identify homologous positions: an example of alignment and data presentation from the frogs. Mol Phylogenet Evol 1995; 4(3): 314-30.
[http://dx.doi.org/10.1006/mpev.1995.1028] [PMID: 8845967]

[25]   Philippe H, Brinkmann H, Lavrov DV, *et al.* Resolving difficult phylogenetic questions: Why more sequences are not enough. PLoS Biol 2011; 9(3)e1000602.
[http://dx.doi.org/10.1371/journal.pbio.1000602] [PMID: 21423652]

[26]   Yli-Mattila T, Paavanen-Huhtala S, Fenton B, Tuovinen T. Species and strain identification of the predatory mite Euseius finlandicus by RAPD-PCR and ITS sequences. Exp Appl Acarol 2000; 24(10-11): 863-80.
[http://dx.doi.org/10.1023/A:1006496423090] [PMID: 11345320]

[27]   Townsend JP. Profiling phylogenetic informativeness. Syst Biol 2007; 56(2): 222-31.
[http://dx.doi.org/10.1080/10635150701311362] [PMID: 17464879]

[28]   Kumazawa Y, Nishida M. Sequence evolution of mitochondrial tRNA genes and deep-branch animal phylogenetics. J Mol Evol 1993; 37(4): 380-98.
[http://dx.doi.org/10.1007/BF00178868] [PMID: 7508516]

[29]   Moritz C, Schneider C, Wake DB. Evolutionary relationships within the *Ensatina eschscholtzii* complex confirm the ring species interpretation. Syst Biol 1992; 41: 273-91.
[http://dx.doi.org/10.1093/sysbio/41.3.273]

[30]   Vanderpoorten A, Goffinet B, Quandt D. Utility of the internal transcribed spacers of the 18S-5.8--26S nuclear ribosomal DNA in land plant systematics with special emphasis on Bryophytes. In: Sharma AK, Sharma A, Quandt D, Eds. Plant Genome: Biodiversity & Evolution. Science Publishers, Enfield NH. 2006; 2B.

[31]   Yodphaka S, Boonpragob K, Lumbsch HT, Kraichak E. Evaluation of six regions for their potential as DNA barcodes in epiphyllous liverworts from Thailand. Appl Plant Sci 2018; 6(8)e01174.
[http://dx.doi.org/10.1002/aps3.1174] [PMID: 30214837]

[32]   Vanderpoorten A, Sotiaux A, Engels P. A GIS based survey for the conservation of bryophytes at the landscape scale. Biol Conserv 2004; 121: 189-94.
[http://dx.doi.org/10.1016/j.biocon.2004.04.018]

[33]   Fisher KM, Wall DP, Yip KL, Mishler BD. Phylogeny of the Calymperaceae with a rank-free systematic treatment. Bryologist 2007; 110(1): 46-73.
[http://dx.doi.org/10.1639/0007-2745(2007)110[46:POTCWA]2.0.CO;2]

[34]   Shaw AJ, Cox CJ, Boles SB. Polarity of peatmoss (Sphagnum) evolution: who says bryophytes have no roots? Am J Bot 2003; 90(12): 1777-87.
[http://dx.doi.org/10.3732/ajb.90.12.1777] [PMID: 21653354]

[35]   Schnittler M, Horn K, Kaufmann R, *et al.* Genetic diversity and hybrid formation in Central European club-mosses (Diphasiastrum, Lycopodiaceae) - New insights from cp microsatellites, two nuclear markers and AFLP. Mol Phylogenet Evol 2019; 131: 181-92.
[http://dx.doi.org/10.1016/j.ympev.2018.11.001] [PMID: 30415022]

[36]   McDaniel SF, Shaw AJ. Selective sweeps and intercontinental migration in the cosmopolitan moss Ceratodon purpureus (Hedw.) Brid. Mol Ecol 2005; 14(4): 1121-32.
[http://dx.doi.org/10.1111/j.1365-294X.2005.02484.x] [PMID: 15773940]

[37]   Fuselier L, Davison PG, Clements M, *et al.* Phylogeographic analyses reveal distinct lineages of the liverworts Metzgeria furcata (L.) Dumort. and Metzgeria conjugata Lindb. (Metzgeriaceae) in Europe and North America. Biol J Linn Soc Lond 2009; 98: 745-56.
[http://dx.doi.org/10.1111/j.1095-8312.2009.01319.x]

[38]   Szövényi P, Hock Z, Schneller JJ, Tóth Z. Multilocus dataset reveals demographic histories of two

peat mosses in Europe. BMC Evol Biol 2007; 7: 144.
[http://dx.doi.org/10.1186/1471-2148-7-144] [PMID: 17714592]

[39]   Shaw AJ, Boles S, Shaw B. A phylogenetic delimitation of the "Sphagnum subsecundum complex"
         (Sphagnaceae, Bryophyta). Am J Bot 2008; 95(6): 731-44.
         [http://dx.doi.org/10.3732/ajb.0800048] [PMID: 21632399]

[40]   Capesius I, Stech M. Molecular relationships within mosses based on 18S rRNA gene sequences.
         Nova Hedwigia 1997; 64: 525-33.

[41]   Cox CJ, Hedderson TAJ. Phylogenetic relationships among the ciliate arthrodontous mosses: evidence
         from chloroplast and nuclear DNA sequences. Plant Syst Evol 1999; 215: 119-39.
         [http://dx.doi.org/10.1007/BF00984651]

[42]   Cox CJ, Goffinet B, Shaw AJ, Boles SB. Phylogenetic relationships among the mosses based on
         heterogeneous Bayesian analysis of multiple genes from multiple genomic compartments. Syst Bot
         2004; 29: 234-50.
         [http://dx.doi.org/10.1600/036364404774195458]

[43]   Davis EC. A molecular phylogeny of leafy liverworts (Jungermanniidae, Marchantiophyta). Monogr
         Syst Bot Missouri Bot Gard 2004; 98: 61-86.

[44]   Forrest LL, Crandall-Stotler BJ. Progress towards a robust phylogeny for the liverworts, with
         particular focus on the simple thalloids. J Hattori Bot Lab 2005; 97: 127-59.

[45]   Hedderson TA, Chapman RL, Cox CJ. Bryophytes and the origins and diversification of land plants:
         new evidence from molecules. In: Bates JW, Ashton NW, Ducketts JG, Eds. Bryologyfor the twenty-
         first century. London: Maney Publishing 1998.

[46]   Forrest LL, Davis EC, Long DG, Crandall-Stotler BJ, Clark A, Hollingsworth ML. Unraveling the
         evolutionary history of the liverworts (Marchantiophyta): multiple taxa, genomes and analyses.
         Bryologist 2006; 109: 303-34.
         [http://dx.doi.org/10.1639/0007-2745(2006)109[303:UTEHOT]2.0.CO;2]

[47]   Budke JM, Goffinet B. Phylogenetic analyses of Timmiaceae (Bryophyta: Musci) based on nuclear
         and chloroplast sequence data. Syst Bot 2006; 31: 633-41.
         [http://dx.doi.org/10.1600/036364406779695861]

[48]   Shaw AJ, Holz I, Cox CJ, Goffinet B. Phylogeny, character evolution, and biogeography of the
         Gondwanic moss family Hypopterygiaceae (Bryophyta). Syst Bot 2008; 33: 21-30.
         [http://dx.doi.org/10.1600/036364408783887311]

[49]   Stech M, Quandt D. 20,000 species and five key markers: the status of molecular bryophyte
         phylogenetics. Phytotaxa 2010; 9: 196-228.
         [http://dx.doi.org/10.11646/phytotaxa.9.1.11]

[50]   Biersma EM, Jackson JA, Stech M, Griffiths H, Linse K, Convey P. Molecular data suggest long-term
         in situ Antarctic persistence within Antarctica's most speciose plant genus, Schistidium. FEVO 2018;
         6: 77.
         [http://dx.doi.org/10.3389/fevo.2018.00077]

[51]   Ochyra R, Smith RIL, Bednarek-Ochyra H. The Illustrated Moss Flora of Antarctica. Cambridge:
         Cambridge University Press 2008.

[52]   Niu JY, Li LL, Shi S, *et al.* Phylogenetic analysis of the genus pohlia (Bryophyta, Bryaceae) using
         chloroplast and nuclear ribosomal DNA. Phytotaxa 2018; 351: 141-53.
         [http://dx.doi.org/10.11646/phytotaxa.351.2.2]

[53]   Liu Y, Medina R, Goffinet B. 350 My of mitochondrial genome stasis in mosses, an early land plant
         lineage. Mol Biol Evol 2014; 31(10): 2586-91.
         [http://dx.doi.org/10.1093/molbev/msu199] [PMID: 24980738]

[54]   Medina R, Johnson M, Liu Y, *et al.* Evolutionary dynamism in bryophytes: Phylogenomic inferences

confirm rapid radiation in the moss family Funariaceae. Mol Phylogenet Evol 2018; 120: 240-7.
[http://dx.doi.org/10.1016/j.ympev.2017.12.002] [PMID: 29222063]

[55]   Sawicki J, Szczecińska M, Bednarek-Ochyra H, Ochyra R. Mitochondrial phylogenomics supports
       splitting the traditionally conceived genus *Racomitrium* (Bryophyta: Grimmiaceae). Nova Hedwigia
       2015; 100: 293-317.
       [http://dx.doi.org/10.1127/nova_hedwigia/2015/0248]

[56]   Szczecińska M, Sramko G, Wołosz K, Sawicki J. Genetic diversity and population structure of the rare
       and endangered plant species *Pulsatilla patens* (L.) mill in east Central Europe. PLoS One 2016;
       11(3)e0151730.
       [http://dx.doi.org/10.1371/journal.pone.0151730] [PMID: 27003296]

[57]   Sawicki J, Szczecińska M, Kulik T, Gomolińska AM, Plášek V. The complete mitochondrial genome
       of the epiphytic moss *Orthotrichum speciosum.* Mitochondrial DNA A DNA Mapp Seq Anal 2016;
       27(3): 1709-10.
       [PMID: 25259451]

[58]   Vigalondo B, Liu Y, Draper I, *et al.* Comparing three complete mitochondrial genomes of the moss
       genus *Orthotrichum* Hedw. Mitochondrial DNA 2016; 1: 168-70.
       [http://dx.doi.org/10.1080/23802359.2016.1149784]

[59]   Sawicki J, Plášek V, Ochyra R, *et al.* Mitogenomic analyses support the recent division of the genus
       Orthotrichum (Orthotrichaceae, Bryophyta). Sci Rep 2017; 7(1): 4408.
       [http://dx.doi.org/10.1038/s41598-017-04833-z] [PMID: 28667304]

[60]   Jonathan Shaw A, Devos N, Liu Y, *et al.* Organellar phylogenomics of an emerging model system:
       Sphagnum (peatmoss). Ann Bot 2016; 118(2): 185-96.
       [http://dx.doi.org/10.1093/aob/mcw086] [PMID: 27268484]

[61]   Groth-Malonek M, Rein T, Wilson R, Groth H, Heinrichs J, Knoop V. Different fates of two
       mitochondrial gene spacers in early land plant evolution. Int J Plant Sci 2007; 168: 709-17.
       [http://dx.doi.org/10.1086/513472]

[62]   Beckert S, Muhle H, Pruchner D, Knoop V. The mitochondrial nad2 gene as a novel marker locus for
       phylogenetic analysis of early land plants: a comparative analysis in mosses. Mol Phylogenet Evol
       2001; 18(1): 117-26.
       [http://dx.doi.org/10.1006/mpev.2000.0868] [PMID: 11161748]

[63]   Quandt D, Bell NE, Stech M. Unravelling the knot: the Pulchrinodaceae fam. nov. (Bryales). Nova
       Hedwigia 2007; 131: 21-39.

[64]   Heslewood MM, Brown EA. A molecular phylogeny of the liverwort family Lepidoziaceae Limpr. in
       Australasia. Plant Syst Evol 2007; 265: 193-219.
       [http://dx.doi.org/10.1007/s00606-006-0512-z]

[65]   Mower JP, Daniel BS, Andrew JA. Plant mitochondrial genome diversity: the genomics revolution. In:
       Jonathan FW, Johann G, Jaroslav D, Ilia JL, Eds. Plant Genome Diversity. Vienna: Springer 2012;
       Vol. 1.
       [http://dx.doi.org/10.1007/978-3-7091-1130-7_9]

[66]   Nickrent DL, Parkinson CL, Palmer JD, Duff RJ. Multigene phylogeny of land plants with special
       reference to bryophytes and the earliest land plants. Mol Biol Evol 2000; 17(12): 1885-95.
       [http://dx.doi.org/10.1093/oxfordjournals.molbev.a026290] [PMID: 11110905]

[67]   Qiu YL, Li L, Wang B, *et al.* The deepest divergences in land plants inferred from phylogenomic
       evidence. Proc Natl Acad Sci USA 2006; 103(42): 15511-6.
       [http://dx.doi.org/10.1073/pnas.0603335103] [PMID: 17030812]

[68]   Jankowiak K, Szweykowska-Kulińska Z. Organellar inheritance in the allopolyploid liverwort species
       Porella baueri (Porellaceae): Reconstructing historical events using DNA analysis data. Monogr Syst
       Bot Missouri Bot Gard 2004; 98: 404-14.

[69] Terasawa K, Odahara M, Kabeya Y, *et al.* The mitochondrial genome of the moss Physcomitrella patens sheds new light on mitochondrial evolution in land plants. Mol Biol Evol 2007; 24(3): 699-709. [http://dx.doi.org/10.1093/molbev/msl198] [PMID: 17175527]

[70] Volkmar U, Knoop V. Introducing intron locus cox1i624 for phylogenetic analyses in Bryophytes: on the issue of Takakia as sister genus to all other extant mosses. J Mol Evol 2010; 70(5): 506-18. [http://dx.doi.org/10.1007/s00239-010-9348-9] [PMID: 20473660]

[71] Gitzendanner MA, Soltis PS, Yi TS, Li DZ, Soltis DE. Plastome phylogenetics: 30 years of inferences into plant evolution. Adv Bot Res 2018; 85: 293-313. [http://dx.doi.org/10.1016/bs.abr.2017.11.016]

[72] Yi YJ, Sun ZW, He S, Sulayman M. A study of molecular sequences, sexuality, and morphological variation in *Plagiomnium carolinianum, P. maximoviczii*, and *P. rhynchophorum* (Bryophyta, Mniaceae). Phytotaxa 2018; 375: 81-91. [http://dx.doi.org/10.11646/phytotaxa.375.1.4]

[73] De-Freitas K, Metz G, Cañon E, Roesch L, Pereira A, Victoria F. Characterization and Phylogenetic Analysis of Chloroplast and Mitochondria Genomes from the Antarctic Polytrichaceae Species *Polytrichum juniperinum* and *Polytrichum strictum.* Diversity (Basel) 2018; 10: 89. [http://dx.doi.org/10.3390/d10030089]

[74] Banting MD, Aquino JDC, David ES, Undan JR. Phylogenetic Analysis of liverworts (Marchantiophyta) in Imugan falls, Santa Fe, Nueva Vizcaya, Philippines using rbcL gene marker. Int J Biol Pharm Allied Sci 2017; 6(1): 89-98.

[75] Aigoin DA, Huttunen S, Ignatov MS, Dirkse GM, Vanderpoorten A. Rhynchostegiella (Brachytheciaceae): Molecular re-circumscription of a convenient taxonomic repository. J Bryol 2009; 31: 213-21. [http://dx.doi.org/10.1179/037366809X12469790518529]

[76] Werner O, Patino J, González-Mancebo JM, Gabriel R, Ros RM. The Taxonomic status and the geographical relationships of the Macaronesian endemic moss Fissidens luisieri (Fissidentaceae) based on DNA sequence data. Bryologist 2009; 112: 315-24. [http://dx.doi.org/10.1639/0007-2745-112.2.315]

[77] Arikawa T, Higuchi M. Preliminary application of psaB sequence data to the phylogenetic analysis of pleurocarpous mosses. Hikobia 2003; 14: 33-4.

[78] Qiu YL, Li L, Wang B, *et al.* A non-flowering land plant phylogeny inferred from nucleotide sequences of seven chloroplast, mitochondrial, and nuclear genes. Int J Plant Sci 2007; 168: 691-708. [http://dx.doi.org/10.1086/513474]

[79] Quandt D, Bell NE, Stech M. Unravelling the knot: The Pulchrinodaceae fam. nov. (Bryales). Nova Hedwigia 2007; 131: 21-39.

[80] Bell NE, Quandt D, O'Brien TJ, Newton AE. Taxonomy and phylogeny in the earliest diverging pleurocarps: square holes and bifurcating pegs. Bryologist 2007; 110: 533-60. [http://dx.doi.org/10.1639/0007-2745(2007)110[533:TAPITE]2.0.CO;2]

[81] Rabeau L, Gradstein SR, Dubuisson JY, Nebel M, Quandt D, Reeb C. New insights into the phylogeny and relationships within the worldwide genus Riccardia (Aneuraceae, Marchantiophytina). Eur J Taxon 2017; 273: 1-26. [http://dx.doi.org/10.5852/ejt.2017.273]

[82] Kijak H, Łodyga W, Odrzykoski IJ. Sequence diversity of two chloroplast genes: rps4 and tRNAGly (UCC), in the liverwort Marchantia polymorpha, an emerging plant model system. History (Lond) 2018; 87: 3573.

[83] Stech M, Sim-Sim M, Esquível MG, *et al.* Phytochemical, molecular, and morphological characterisation of the liverwort genus Radula in Portugal (mainland, Madeira, Azores). Syst Biodivers 2010; 8: 257-68.

[http://dx.doi.org/10.1080/14772001003723579]

[84] Köckinger H, Lüth M, Werner O, Ros RM. Tortella mediterranea (Pottiaceae), a new species from southern Europe, its molecular affinities, and taxonomic notes on T. nitida. Bryologist 2018; 121: 560-70.
[http://dx.doi.org/10.1639/0007-2745-121.4.560]

[85] Patwardhan A, Ray S, Roy A. Molecular Markers in Phylogenetic Studies – A Review. J Phylogenetics Evol Biol 2014; 2: 131.

[86] Stech M, Frey W. A morpho-molecular classification of the mosses (Bryophyta). Nova Hedwigia 2008; 85: 1-21.
[http://dx.doi.org/10.1127/0029-5035/2008/0086-0001]

[87] Devos N, Vanderpoorten A. Range disjunctions, speciation, and morphological transformation rates in the liverwort genus Leptoscyphus. Evolution 2009; 63(3): 779-92.
[http://dx.doi.org/10.1111/j.1558-5646.2008.00567.x] [PMID: 19154356]

[88] Long DG, Möller M, Preston J. Phylogenetic relationships of *Asterella* (Aytoniaceae, Marchantiopsida) inferred from chloroplast DNA sequences. Bryologist 2000; 103: 625-44.
[http://dx.doi.org/10.1639/0007-2745(2000)103[0625:PROAAM]2.0.CO;2]

[89] Wynns JT, Schröck C. Range extensions for the rare moss *Plagiothecium handelii*, and its transfer to the resurrected genus *Ortholimnobium*. Lindbergia 2018; 41: 01087.
[http://dx.doi.org/10.25227/linbg.01087]

CHAPTER 5

An Overview of the Family Calymperaceae (Bryophyta) in Western Ghats with Special Reference to Kerala and Its Status in India

Manju C. Nair*, V. K. Chandini and K. P. Rajesh

PG & Research Department of Botany, The Zamorin's Guruvayurappan College, Kozhikode-14, Kerala, India (affiliated to the University of Calicut)

Abstract: The Calymperaceae occurring in Kerala are described here in. Five genera are recognised within the family *viz.*, *Calymperes* Sw., *Exostratum* L. T. Ellis, *Leucophanes* Brid., *Octoblepharum* Hedw. and *Syrrhopodon* Schwägr. In Kerala, *Calymperes* is represented by eight species, *Syrrhopodon* by five, *Leucophanes* by *L. glaucum* and *L. octoblepharioides*, *Exostratum* by *E. blumii*, and *Octoblepharum* by *O. albidum* Hedw., *Exostratum* is a new genus record and *Syrrhopodon prolifer* is a new species record for Kerala.

Keywords: Calymperaceae, Distribution, India, Kerala, Taxonomy, Western Ghats.

INTRODUCTION

The family Calymperaceae have primarily tropical and subtropical distribution, but a few species reach temperate latitudes [1 - 3] Most members are epiphytic and occur at low to moderate elevations. Kindberg [4] was the first to recognise the Calymperaceae as a family, although his concept included only *Calymperes*. He classified *Syrrhopodon* within the Weisiaceae [=Pottiaceae].

Andrews [5] considered parts of the Leucobryaceae and the Calymperaceae to be closely allied. Based on the structure of the peristome and the anatomy of the leaf, he proposed the elimination of the family Leucobryaceae, placing *Leucophanes* Brid., *Arthrocormus* Dozy and Molk., *Exodictyon* Card. and *Octoblepharum* Hedw. in the Calymperaceae, but including *Leucobryum* Hampe and related genera in the Dicranaceae.

---

* **Corresponding author Manju C. Nair:** PG & Research Department of Botany, The Zamorin's Guruvayurappan College, Kozhikode-14, Kerala, India (affiliated to the University of Calicut); Tel: +91-9447439634; Email: manjucali@gmail.com

Afroz Alam (Ed.)

In a recent synthesis of the various opinions, Buck and Goffinet [6] included *Calymperes* Sw. ex F. Weber, *Syrrhopodon* Schwägr., *Mitthyridium* H. Rob., *Chameleion* L. T. Ellis & A. Eddy, *Arthrocormus* Dozy & Molk., *Exostratum* L. T. Ellis, *Exodictyon* Cardot, *Leucophanes* Brid. and *Octoblepharum* Hedw. in Calymperaceae.

Calymperes Sw. ex F. Weber, *Syrrhopodon* Schwägr., *Octoblepharum* Hedw., *Exostratum* L. T. Ellis, *Leucophanes* Brid. and *Mitthyridium* H. Rob. (≡*Thyridium* Mitt., *nom. illeg.*) are represented in India [7]. But Gangulee [7] treated *Exodictyon* (synonym of *Exostratum*), *Octoblepharum* Hedw., *Ochrobryum* Mitt., *Leucobryum* Hampe and *Leucophanes* Brid. under Leucobryaceae and *Calymperes, Syrrhopodon* and *Thyridium* (*Mitthyridium*) under Calymperaceae. *Exostratum* L. T. Ellis is a new genus record of occurrence to Kerala.

Calymperes Sw. ex F. Weber includes about 54 species worldwide [8], many of which occupy broad ranges in the palaeo-or neotropics. A few species, including *C. afzelii* Sw., *C. erosum* Müll. Hal., *C. palisotii* Schwägr., *C. lonchophyllum* Schwägr. and *C. tenerum* Müll. Hal. are pantropical [9]. In India, 17 species and one variety of the genus *Calymperes* are reported by the study [7]. Among them, the validity of *C. noakhalensis* Bruehl and Sarkar and *C. tenerum* var. *teniolata* Gangulee has yet to be established, but all other taxa described by Gangulee [7] were synonymised under different species [10 - 13]. *C. burmense* Hampe, *C. manii* Müll. Hal. ex Besch., *C. hampei* Dozy and Molk., *C. heterophyllum* Hampe = *C. erosum* Müll. Hal.; *C. kurzianum* Hampe ex Müll. Hal., *C. calcuttense* E. B. Bartram and Gangulee = *C. moluccense* Schwägr.; *C. andamense* Besch. = *C. taitense* (Sull.) Mitt.; *C. vriesei* Besch. = *C. afzelii* Sw.; *C. lingulatum, C. nicobarense* Hampe, *C. punctulatum* Hampe = *C. graeffeanum* Müll. Hal.; *C. delessertii* Besch. = *C. boulayi* Besch. *C. sikkimense* Broth., *C. sundarbanense* Gangulee, *C. griffithii* Müll. Hal. = *Chameleion peguense* (Besch.) L. T. Ellis and A. Eddy [14]. The presence in India of two other species, *Calymperes tenerum* Müll. Hal. and *C. mangalorense* Dixon and P.de la Varde, was confirmed by Ellis [12]. Here record 14 valid species for India/Sri Lanka. In Kerala this genus is represented by, *C. erosum, C. graeffeanum* Müll.Hal., *C. taitense* (Sull.) Mitt., *C. afzelii* Sw., *C. lonchophyllum* subsp. *lonchophyllum, C. tenerum* Müll. Hal. and *C. mangalorense* Dixon and P. de la Varde.

Chameleion L.T. Ellis and A. Eddy is a tropical genus and the name was proposed by Ellis and Eddy [15], worldwide it includes three species *viz. C. peguense* (Besch.) L. T. Ellis and A. Eddy, *C. cryptocarpus* and *C. xanthophyllus*. Only *C. peguense* (Besch.) L. T. Ellis and A. Eddy occurs in India, have been described under several different names, such as *Calymperes sikkimense* Broth., *C. sundarbanense* Gangulee and *C. griffithii* Müll. Hal. *Chameleion peguense* (Besch.) L. T. Ellis and A. Eddy was recorded from Kerala [16].

Exostratum L. T. Ellis has an Indo-Pacific distribution and includes four species. Only *E. blumii* (Nees ex Hampe) L. T. Ellis is found in India, reported as '*Exodictyon blumii*' [7]. It also occurs in Kerala.

Octoblepharum Hedw. is widely distributed, but best represented in the neotropics. In the Old World tropics, the pantropical *O. albidum* Hedw. is the most commonly occurring species. It is widely distributed from low to high altitudes in Kerala. *O. arthrocormoides* N. Salazar and B. C. Tan, recently described from Tropical Asia [17] is yet to be found in India.

Leucophanes Brid. shows two centres of species diversity, one in Africa, and the other in Malaysia and adjacent areas. Gangulee [7] recorded four species of *Leucophanes* Brid. from India *viz., L. glaucescens* Müll. Hal. ex M. Fleisch., *L. albescens* Müll. Hal., *L. octoblepharoides* Brid. and *L. nicobaricum* Müll. Hal. *L. albescens* Müll. Hal. is synonymised under *L. glaucum* (Schwägr.) Mitt. and *L. nicobaricum* Müll. Hal. under *L. octoblepharoides* Brid., *L. glaucum* (Schwägr.) Mitt. and *L. octoblepharoides* Brid. have been reported in Kerala [18]. All of these species are distributed at high altitudes.

Syrrhopodon is a pantropical genus with approximately 116 species worldwide [8]. Gangulee [7] recorded *S. rufescens* Hook. and Grev., *S. larminatii* Broth. and Paris, *S. spiculosus* Hook. and Grev., *S. gardneri* (Hook.) Schwaegr., *S. prolifer* Schwägr., *S. assamicus* H. Rob and *S. subconfertus* Broth. in India, *S. parasiticus* (Brid.) Besch. was reported by Manju *et. al.* [19]. *Syrrhopodon gardneri* (Hook.) Schwaegr., *S. parasiticus* (Brid.) Besch., *S. prolifer* Schwägr., *S. rufescens* Hook. and Grev. and *S. spiculosus* Hook. and Grev. are valid species. Three species are synonymised under different names *viz., Syrrhopodon assamicus* H. Rob. = *S. semiliber* (Mitt.) Besch [20]; *S. larminatii* Broth. and Paris = *S. armatus* Mitt.; *S. subconfertus* Broth. = *S. confertus* Sande Lac [21]. In Kerala this genus is distributed with five species *viz., S. gardneri* (Hook.) Schwaegr., *S. prolifer* Schwägr., *S. spiculosus* Hook. and Grev., *S. parasiticus* (Brid.) Besch. and *S. rufescens* Hook. and Grev.

The present study describes the members of the Calymperaceae present in Western Ghats with a special reference to Kerala.

MATERIALS AND METHODS

The specimens for the present study were collected from 2004-2018 from low to high altitude areas in different seasons. A key to the genera and species is provided for easy identification. The species is identified using authentic literatures and consulting with experts. Each species is provided with author citation, synonyms, description, habitat, distribution and specimens examined. Some abbreviations used in the text are MCN; Manju C.N, KPR; K.P. Rajesh, PVM; P.V. Madhusoodanan and SVK, Sreenivas V.K. The specimens collected were deposited in the Calicut University Herbarium (CALI) and The Zamorin's Guruvayurappan College herbarium (ZGC).

RESULTS

Key to the Genera of Calymperaceae in Kerala

1a. Costa with dorsal and ventral bands or steroids separated by 1-2 or more layers of guide cells 2
1b. Costa with large empty hyaline cells supporting network or layers of small chlorophyllose cells 4
2a. Peristome haplolepidous, calyptra fugacious 3
2b. Peristome absent, calyptra persistent .... *Calymperes*
3a. Leaves never dimorphic, border between hyaline and chlorophyllose lamina sharply defined, superficial cells with smooth or with teeth, spines or coronate papillate projections ....*Syrrhopodon*
3b. Leaves sometimes dimorphic, border between hyaline and chlorophyllose lamina poorly defined, superficial cells ventrally protuberant, smooth or papillose .... *Chameleion*
4a. Plants whitish, without central strand 5
4b. Plants greenish, with central strand .... *Exosratum*
5a. Cells near the margin, rectangular .... *Octoblepharum*
5b. Cells near margin hyaline, 3-4 layers elongated cells from base to apex ......
....*Leucophanes*

Key to Species of *Calymperes*

1a. Leaves mostly long linear to long-subulate or acuminate 2
1b. Leaves variously shaped 3
2a. Cancellinae sharply demarcated distally from adjacent cells of upper lamina, cancellinae often excavate .... *C. lonchophyllum*
2b. Cancellinae gradually merging with adjacent cells of upper lamina distally, cancellinae not excavate 4
3a. Non-gemmiferous leaves lingulate with obtuse-apiculate or denticulate apices; costa never excurrent ....*C. afzelii*
3b. Non-gemmiferous leaves lingulate to narrowly lanceolate with acute, denticulate apices; costa always excurrent, smooth at apex .... *C. erosum*
4a. Costa shortly excurrent in gemmiferous leaves and bearing gemmae all around on its tip... *C. tenerum*

4b. Costa at most percurrent in gemmiferous leaves, bearing gemmae only on ventral surface at its tip 5

5a. Leaves often coarsely toothed above, teeth in two rows .... *C. taitense*

5b. Leaves entire or weakly toothed above, teeth in single row 6

6a. Gemmiferous leaves sharply differentiated from vegetative leaves, gemmae only on adaxial surface, costa percurrent 7

6b. Gemmiferous leaves differentiated only at their tips, but similar to vegetative leaves; gemmae seen at tip radiating from the centre, costa excurrent .... *C. mangalorense*

7a. Cells of chlorophyllose lamina 4-7 μm wide, protruding roundly to subacutely from the ventral leaf surface, apices of gemmifrous leaves rounded or truncate ....*C. graeffeanum*

7b. Cells of chlorophyllose lamina 6-11 μm wide, protruding roundly from the ventral leaf surface, apices of gemmiferous leaves rounded or obtusely pointed ....*C. palisotii*

Calymperes afzelii Sw., Jahrb. Gewachsk. 1: 3. 1818. *C. thwaitesii* Besch., Annl. Sci. Nat. Bot. ser. 8, 1: 306. 1895. *C. vriesei* Besch., Ann. Sci. Nat. Bot. Sr 8,1: 268, 307. 1895. *C. microdictyon* Dixon & P.Varde, Ann. Crypt. Exot. 3: 172. 1930.

Plants yellowish-green, 3 cm or more high, in tufts; leaves incurled when dry, erect when moist, leaves up to 5 mm long, dimorphic; non-gemmiferous leaves lingulate with obtuse or obtuse-apiculate or denticulate apices; costa never excurrent; cells above leaf base quadrate to shortly rectangular, basal cells hyaline, hyaline lamina sharply defined, quadrate to rectangular, marginal lamina at base entire; cells of chlorophyllose lamina mostly isodiametric or slightly longer than broad with 4-6 sides, cells of distal hyaline lamina mostly isodiametric, leaf margin from above hyaline base to near apex consisting of strong polystratose marginal rib; gemmiferous leaves lanceolate to lingulate with a narrow, gemmae bearing proboscis; costa strong, extending into proboscis but ending below apex, lamina narrowing into proboscis, near the leaf apex forming rounded, dentate tip; gemmae arising in a radial mass from the ventral surface of the costal apex, fusiform, smooth, often red. (Fig. **1 A-H**)

Fig. (1). A-H: *Calymperes afzelii,* A. Habit, B. Leaf, C. Leaf Base, D. Gemmiferous leaf tip, E-F. Submarginal cells at base, G. Leaf basal cells at costa, H. Spinulose cells at costa (from MCN & KPR 106663).

Habitat & Distribution

Commonly occurs on shaded tree trunks and logs including rotting stumps and fern rhizomes in moist deciduous and evergreen forests. A pantropical species. It is a common species in the study area, being widely distributed in South India (Goa, Kerala, Tamil Nadu), Sri Lanka and China [12, 7 as *C. vriesei*).

Specimens Examined

Kerala, Wayanad WLS, Mananthavady (700 m) *MCN, PVM & KPR 80113c,* Kurichiad range (880 m) *MCN 84558a*; Thiruvananthapuram, Near *Agasthyamalai* peak (1500 m) *MCN & KPR 106663;* Kozhikode, Kakkayam, Valayamchal (150 m) *MCN 87605* (CALI), Kakkavayal (100 m), *Reshma, MCN & KPR 1076b* as *C. vriesi* (ZGC).

Calymperes erosum Müll.Hal., Linnaea 21: 182. 1848. *Syrrhopodon heterophyllus* Mitt., J. Linn. Soc. Bot. Suppl. 1: 40.1859. *Calymperes burmense* Hampe in Besch., Ann. Sci. Nat. Bot. ser. 8, 1: 267, 279. 1895. *C. fordii* Besch., Ann. Sci. Nat. Bot. ser. 8, 1: 267, 284. 1895. *C. exlimbatum* Müll.Hal., ex Besch., Ann. Sci. Nat. Bot. ser. 8, 1: 267. 1895. *C. manii* Müll.Hal. ex Besch., Ann. Sci. Nat. Bot. ser. 8, 1: 267, 291. 1895. *C. seychellarum* Besch., Ann. Sci. Nat. Bot. Ser. 8, 1: 267, 304. 1895. *C. thyridioides* Müll.Hal. *in* Paris, Index Bryol. Suppl. 86. 1900. *C. subcrassilimbatum* P. Varde, Svensk Bot. Tidskr., 51: 159. 1957.

Plants greenish, 1-4 cm high, in tufts; leaves incurled when dry, erecto-patent when moist, mostly 3-5 mm long, dimorphic, non-gemmiferous leaves lingulate to narrowly lanceolate with acute, denticulate apices; costa always excurrent, cells of chlorophyllose lamina mostly isodiametric to slightly longer than broad, rounded-hexagonal, protruding acutely from the ventral leaf surface, dorsally smooth or with one or two papillae; teniola continuous from leaf base to a short distance below the apex, base with well defined, intramarginal unistratose band of long, narrow thick walled cells, upper hyaline base cells becoming shorter, forms denticulate margin, at and above shoulders cells are small and thick walled; gemmiferous leaves narrowly lanceolate, similar to non-gemmiferous leaves; costa strongly excurrent, usually broadening slightly towards the tip, gemmae produced in radial mass from all around the costal apex, pale to dark brownish to greenish-brown (Fig. **2 A-N**).

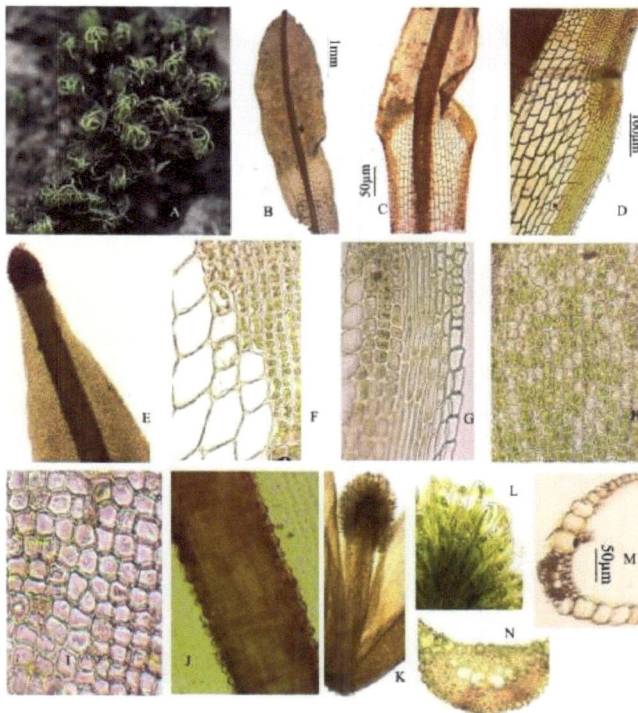

Fig. (2). A-N. *Calymperes erosum*: A. Habit, B. Leaf, C. Leaf at sheathing base, D. Leaf basal margin, J. Costa, H-L. Gemmae, M-N. Cross section of leaf (D-L. Same size; from MCN 87449/a:CALI).

Habitat & Distribution

Usually epiphytic, often associated with *Octoblepharum albidum* and *Lejeunea* sp., on logs and as epiphylls. Pantropical species. It is a common species in South India (Goa, Kerala, Tamil Nadu), Sri Lanka, Myanmar, China, Africa and America [12,7 as *C. burmense* and *C. manii*).

Specimens Examined

Kerala, Wayanad, Vythiri (750 m) *MCN & PVM 80099*; Kannur, Aralam WLS, Chavachi (180 m), *MCN 87549/a* (CALI); Kozhikode, Kakkavayal (100 m), *Reshma, MCN & KPR 1062, 1175a* (ZGC); Thrissur, Peechi- Vazhani Wildlife sanctuary, Vellani, Chooralkkunnu (90m) *Chandini & Manjula 11362* (ZGC).

Calymperes graeffeanum Müll.Hal., J.Mus. Godeffroy. 3(6): 64. 1874.

Plants small, pale-greenish, in small turfs. Stem 1-2 cm long, rhizoids reddish brown. Leaves, curled and contorted when dry, straight when wet, dimorphic: gemmiferous leaves sharply differentiated from non-gemmiferous leaves, non-gemmiferous leaves oblong-lanceolate, gemmiferous leaves acuminate distally with often spatulate tips, 2.5 - 4.0 mm long, costa percurrent, often long-extended beyond broad lamina, above hyaline base of costa rough with small acute projection; cells of chlorophyllose lamina 4-7 μm wide, protruding roundly to subacutely from the ventral leaf surface, isodiametric, obscure, smooth to unipapillose abaxially, mammillose-papillose adaxially; margin thickened (polystratose), margin of leaf base entire, sometimes irregularly serrate at leaf apex; from leaf base to below apex of hyaline lamina with a narrow intramarginal band of thick walled cells; hyaline lamina usually broadly to narrowly acute distally. Gemmiferous leaves lanceolate to narrowly lingulate with an apical gemma bearing proboscis, apex rounded to truncate, costa thick ending below apex, gemmae common, borne at leaf apex on adaxial surface of costa.

Habitat & Distribution

On land cuttings and on bark. This species was reported from Kerala as *C. punctulatum* [22]. The present collection forms the authentic report of the occurrence of this species in Kerala. It has also been reported from Tamil Nadu and the Andaman Islands; China, Sri Lanka, Philippines, Australia and Reunion Island [7, 12, 14].

Specimens Examined

Kerala, Kannur, Aralam WLS, Uruppukunnu (750 m) *KPR 99721*; *KPR 106500, 106517 (CALI)*; Thrissur, Peechi-vazhani Wildlife sanctuary, Peechi dam area (82 m), *Chandini & Mufeed 10490 (ZGC)*.

Calymperes lonchophyllum Schwägr., Sp. Musc. Frond., Suppl. 1. 2:333. 1816.

Plants dark greenish, unbranched, in low tangled tufts, 1-2.2 cm high. Stems very short, often almost lacking; rhizoids glossy blackish red. Leaves monomorphic, curled when dry, straight when wet, linear above slightly broader base, to 1 cm long, with abundant pale paraphysis-like axillary hairs with longest median cells 3-4 or more times as long as wide, costa ending in leaf apex to short excurrent; chlorophyllose lamina mostly unistratose occasionally bistratose, mostly or partly transversely elongate, smooth to papillose abaxially and adaxially, cells in surface view rounded, irregularly polygonal to shortly rectangular, margins thickened, polystratose, toothed in two rows, the teeth remote in proximal part, hyaline lamina sharply defined; leaf margin in base with an intramarginal rib of thick walled linear cells; Gemmae infrequent, adaxial on leaf tips, yellowish brown. (Fig. **3 A-I**)

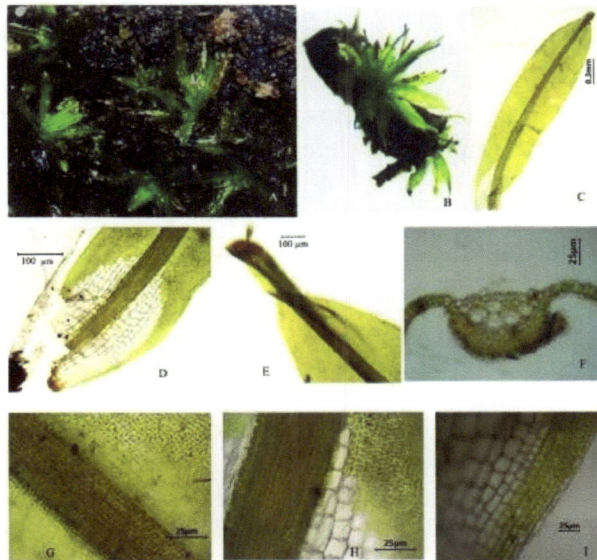

Fig. (3). *Calympere lonchophyllum*: A&B. Habit, C. Gemmiferous leaf, D. Leaf base, E. Leaf tip with gemmae, F. Cross section of leaf, G. Leaf middle cells, H. Hyaline cells at base, I. I. Leaf basal marginal cells (ZGC:702).

Habitat & Distribution

On tree trunks and rocks in evergreen forests at moderate elevations; A widely distributed pantropical species [14].

Specimens Examined

Kerala, Thiruvananthapuram, Agasthyamalai Biosphere Reserve, Way to *Pandavamala* (1350 m) *SVK 106637* (CALI); Malappuram, New Amarambalam Reserve Forest, Nilambur, *KPR 702* (ZGC).

Calymperes mangalorense Dixon & P. de la Varde, Arch. Bot. Bull. Mens. 1(8–9): 164. 1927.

Plants upto 2 cm long, yellowish green; leaves 4 mm long, dimorphic but not markedly so, non-gemmiferous leaves narrowly to broadly lingulate, distal leaves involute, leaf apex acute to obtuse, costa ends at apex to excurrent, above hyaline base rough with small acute projections, cells of chlorophyllose lamina subisodiametric, protruding acutely from the ventral leaf surface, dorsally flat to unipapillose; leaf margin beyond hyaline lamina to apex uneven, polystratose, composed of subquadrate to shortly subrectangular chlorophyllose cells, from distal hyaline lamina to leaf base with an intramarginal unistratose band of long rectangular to linear thick walled cells, marginal lamina consisting of 1-3 rows of thin walled, subrectangular hyaline cells, entire to denticulate; gemmiferous leaves differentiated only at their tips, lanceolate to lingulate, but similar to vegetative leaves. In gemmiferous leaves, costa excurrent, gemmae produced from all around the costal tip. (Fig. **4 A-I**).

Habitat & Distribution

Epiphytic, Endemic to India, distributed in Mangalore, Goa (Ellis, 1989) and Kerala. Ellis (1989) commented that the protologue of *C. tenerum* var. *teniolatum* described by Gangulee [7], resembles *C. mangalorense*, hence its distribution ranges to Calcutta also.

Specimen Examined

Kerala, Kozhikode, Ramanattukara (sea level) *MCN & KPR 6367* (ZGC).

Calymperes palisotii Schwägr., Sp. Musc. Frond., Suppl. 1 2: 334, pl. 98. 1816.

Fig. (4). *Calymperes mangalorense*: A. Habit, B. Cells near to leaf margin-KOH, C. Single leaf, D. Leaf base, E. Leaf base (marginal), F. Cross section of leaf base-KOH, G. Marginal cells of leaf base, H. Apical portion of leaf (10X), I. Apical-marginal portion of leaf (from ZGC 6367).

Plants 0.3 mm to 2 cm long, yellowish green, leaves incurled when dry, erect when moist, 1.5-2.8 mm long, dimorphic; non gemmiferous leaves lingulate to narrowly obovate with obtusely pointed apices, sometimes apiculate, costa ending in or just below apex, ventral superficial cells above leaf base subquadrate to shortly rectangular, often wider than long, protruding roundly from the costa surface, dorsal superficial cells long, narrowly subrectangular, toward apex similar to ventral cells; cells of chlorophyllose lamina subquadrate to subhexagonal, protruding roundly from the ventral leaf surface, 6-11 μm, dorsally flat or unipapillose; cells of hyaline lamina with more or less uniformly thin walls, teniola extending from leaf base to a short distance below the apex, around shoulders of leaf interrupted by the chlorophyllose lamina; below shoulders forming a well-defined intramarginal unistratose band, three to six cells wide, cells linear, thick walled; above shoulders consisting of a narrow polystratose rib, and forming an entire leaf margin; gemmiferous leaves lanceolate to lingulate with modified apices, proboscis suboblong; costa strong, extending into proboscis, ending below leaf apex, lamina sharply narrowing into proboscis and becoming tightly recurved, broadening slightly above and becoming plane to form a narrow margin around the tip of the costa, ending as an obtusely pointed or rounded leaf apex; gemmae frequent arising from ventral surface of costal apex (Fig. **5 A-H**).

Fig. (5). *Claymperes palisotti*: A. Habit, B. Single leaf, C. Leaf apex, D. Leaf apex (enlarged), E. Leaf base, F. Leaf margin, G-H. Cross section of leaf.

Habitat & Distribution

Papua New Guinea, Panama, Nigeria, Senegal, Tanzania (O' Shea, 2006); India, Sri Lanka, Maldives. (Description and distribution after [12].

Specimen Examined

Kerala, Kottayam, Changanacherry, 1931, C. John 40 (BM), we could not locate this species at this time.

Calymperes taitense (Sull.) Mitt., J. Linn. Soc., Bot. 10: 172. 1868. *Syrrhopodon taitense* Sull., U.S. Expl. Exp. Wilkes Musci 6. 1860.

Plants dark greenish, blackish proximally, in tufts and cushions. Stem elongate, 2-5.5 cm long, often branched, rhizoids glossy, dark reddish. Leaves monomorphic, curled and contorted when dry, straight and erect-spreading when wet, oblong-lanceolate to broadly linear above broader at base, 5-7 mm long; costa extending into proboscis, with narrow, revolute, lateral wings of lamina; chlorophyllous lamina abruptly narrowing into proboscis and becoming narrowly recurved, apex dentate, cells of upper lamina isodiametric, small, obscure, smooth abaxially, bulging adaxially, margins thickened distally (polystratose), toothed proximally, coarsely toothed distally, hyaline lamina sharply defined; from proximal limit of marginal rib to leaf base with a narrow unistratose to polystratose intramargnal band of linear thick walled cells. Gemmae numerous, borne adaxially on modified club-like tip of costa (Fig. **6 A-H**).

Habitat & Distribution

Epiphytic, found on dried twig; South India (Kerala), Andaman Islands, China [12]. This species was earlier reported as *Calymperes andamense* from Kerala [23]. A paleotropical species reaching the Pacific Island.

Specimen Examined

Kerala, Kozhikode, Kakkavayal (100 m), *Reshma, MCN & KPR 1055* (ZGC).

Calymperes tenerum Müll.Hal., Linnaea. 37:174. 1872.

Plants small, pale, compact, in low turfts. Stem short, branched; rhizoids brownish. Leaves hardly dimorphic, 2–3 mm long, not much contorted when dry, usually secund, straight and erect-spreading when wet, vegetative leaves oblong, gemmiferous leaves oblong or acuminate; costa usually excurrent in all leaves; cells of upper lamina isodiametric, smooth to unipapillose abaxially, mammillose-papillose adaxially; margins unistratose or slightly thickened, entire throughout;

teniolae lacking; hyaline lamina narrow, typically truncate distally. Gemmae common, in pale spherical clusters all around on tip of gemmiferous leaves. Sporophytes not seen (Fig. **7 A-O**).

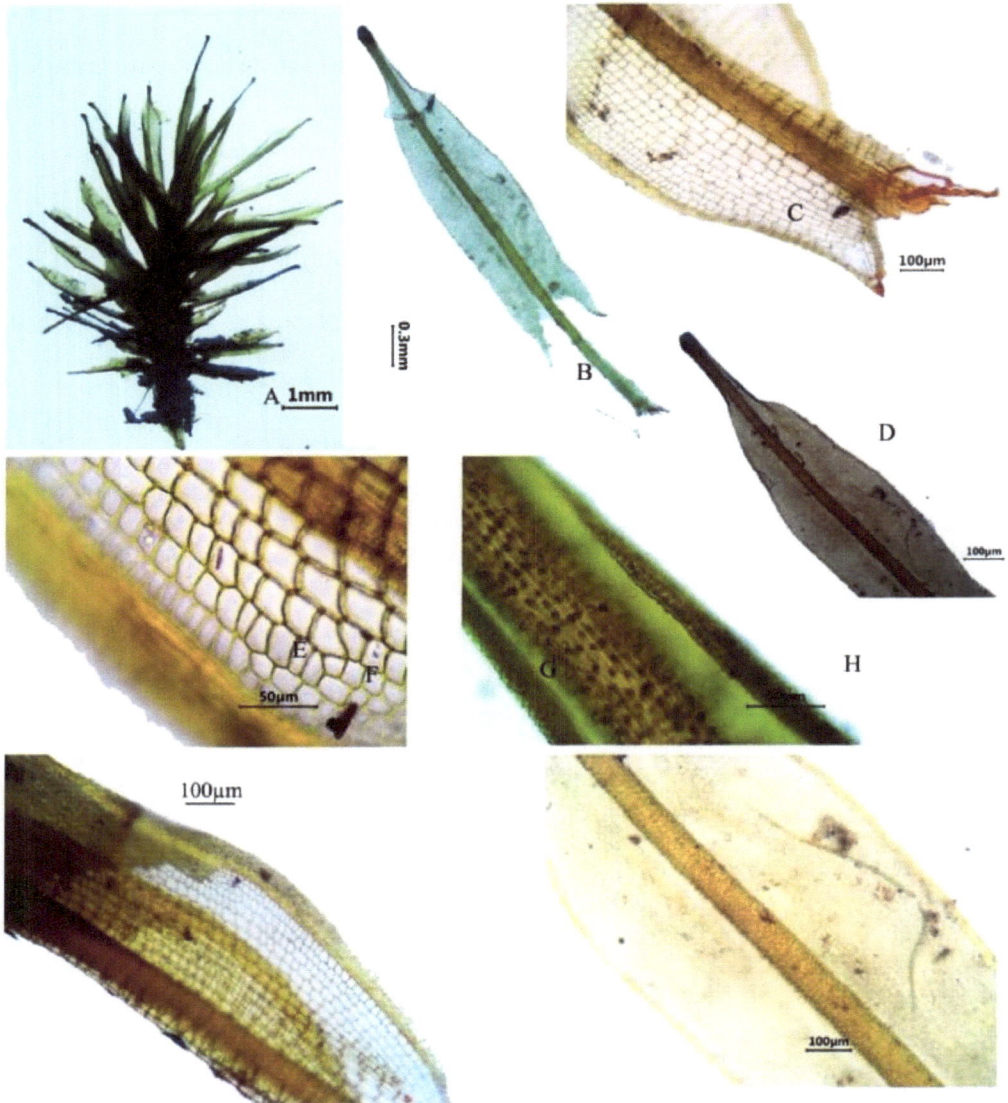

Fig. (6). A-H: *Calymperes taitense*: A. Habit, B. Leaf, C. Leaf base, D. Gemmiferous leaf tip, E-F. Submarginal cells at base, G. Basal cells of leaf at costa, H. Spinulose cells at costa (from MCN & KPR: 106663).

Fig. (7). A-O. *Calymperes tenerum*: A. habit, B. Single plant, C-D. Gemmiferous leaf, E. Gemmiferous leaf tip, F. Costa, G. Leaf base, H. Leaf cells (upper middle), I. Leaf margin at the emergence of teniola, J. Marginal cells of leaf tip, K. Extere basal cells of leaf, L. Gemmae, M. Cross section of leaf base, N. Cross section of leaf (basal), O. Antheridia with paraphyses.

Habitat & Distribution

It is found on tree trunks in humid forests at low elevations, often very common and it is abundant in much of the palaeotropics but very rare in the neotropics [3]. It is distributed in Northern India (Bengal), Southern India (Tamil Nadu, Kerala), Sumatra, Malay, Java, Molucca, New Guinea, Philippines, New Caledonia, Hawaii, Caroline Is., Samoa and Tahiti [12, 14].

Specimens Examined

Kozhikode, Kakkayam, Valayamchal (40 m), *MCN & KPR 99705* (CALI); Thrissur, Thrissur, Peechi- Vazhani Wildlife sanctuary, Vaniyampara, Aukkalapara (90m) *Chandini 11449* (ZGC).

Chameleion peguense (Besch.) L. T. Ellis & A. Eddy in A. Eddy, Handb. Males. Mosses 2: 250. 1990. *Calymperes peguense* Besch., Ann. Sci. Nat. Bot. ser. 8,1: 269, 299: 1895. *Heliconema peguense* L.T. Ellis & A. Eddy, J. Bryol. 15: 730. 1989, *hom. Illeg. Syrrhopodon peguense* (Besch.) W. D. Reese, J. Hattori Bot. Lab. 82: 243. 1997.

Plants 0.5-1.2 mm long, forming mats; leaves dimorphic, nongemmiferous leaves erect, 3 mm long, narrowly to broadly lingulate, distal lamina incurved, apex broadly obtuse to rounded, entire; costa ends just below apex, ventral cells above the hyaline base shortly rectangular, slightly protuberant, dorsal cells longer with a simple papilla near end walls; cells of chlorophyllose lamina isodiametric to slightly longer than wide, mostly 5-10µm; hyaline lamina not sharply defined, cells near the costa thick walled, brownish, superficial walls with transverse bands of thickening; leaf margin above hyaline base irregularly and distantly toothed, unistratose or incorporating a narrow intramarginal polystratose rib, which dorsally may give rise to low lamellae; gemmiferous leaves linear, longer that non gemmiferous leaves, 5 mm long, costa strong, ending below apex, distally rough with many superficial cells drawn out as acute projections, chlorophyllose lamina usually involute, cells subquadrate to long subrectangular, hyaline lamina narrow with cells similar to those in nongemmiferous leaves. Gemmae produced from the ventral surface of the costal apex. (Fig. **8 A-F**)

Habitat & Distribution

It is found on trunks, branches and exposed roots of trees. A rare species recorded from India, Indochina, Malaysia, Myanmar (Burma), Vietnam and the Philippines [14, 24].

Specimen Examined

Kerala, Wayanad, Soochippara (850 m) *KPR 87057 (*CALI).

Exostratum blumii (Nees ex Hampe) L.T.Ellis, Lindbergia 11: 25.1985. *Syrrhopodon blumei* Nees ex Hampe, Bot. Zeitung (Berlin) 5: 921 (1874). *Exodictyon blumei* (Nees ex Hampe) M.Fleisch., Musci Frond. Archip. Ind. Exsicc. 58 (1899). *Exodictyon subscabrum* (Broth.) Cardot, Rev. Bryol. 26: 7 (1899). H.C.Gangulee, 1972. *op. cit.* 445, fig. 208, as *Exodictyon blumei*.

Plants slender, forming thick pale whitish-green tufts. Stem erect, with numerous rhizoids from leaf axils, 1-3 cm long. Leaves 3–4 mm long, erect to patent, ascending from an erect sheathing base as a long narrowly lanceolate or linear subula, apex acute, dentate, leaf margin serrulate, at shoulder strongly serrulate. Leaf bordered from base to apex by 2 or 4 rows of very elongated narrow incrassate cells; leaf base composed of long, narrow hyalocysts, distally becoming shorter, leaf margin in subula formed by a rib of long narrow hyalocysts enclosed by chlorocysts, in the leaf base formed by a flat band of linear cells. Costa composed of 4–8 layers of hyalocysts enclosed in a cortex of chlorocysts, and with a median band of chlorocysts, superficial chlorocysts mostly with apically directed spines; gemmae frequent, produced at apex of costa, fusiform, uniseriate (Fig. **9 A-K**).

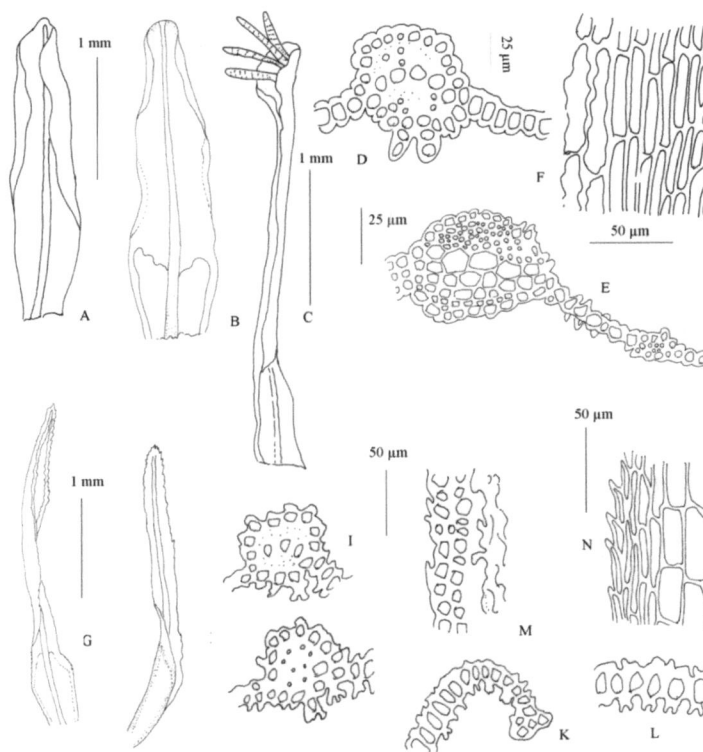

Fig. (8). A-F. *Chameleion peguense*: A-B. Non gemmiferous leaves, C. Gemmiferous leaf, D. Costa of gemmiferous leaf, E. Costa of non-gemmiferous leaf, F. Margin at hyaline base, G-N. *Syrrhopodon gardenerii*: G-H. Ventral leaves, I-J. Section through costa, K-L. Chlorophyllous lamina, M. Marginal cells of chlorophyllous lamina, N. Marginal cells of chlorophyllous lamina.

Fig. (9). A-K. *Exostratum blumii*: A. habit, B, C, K. Leaves, D. Leaf apex, E-J. Leaf margin at middle, F. Leaf margin at neck, G. Cross section of leaf, H. Basal cells of leaf, I. Basal cells of leaf (enlarged); L-X. *Octoblepharum albidum*: L. Habit, M. Leaf, N, V, W. Leaf bases, O. Leaf apex, P. Leaf tip (enlarged), Q. Basal cells of leaf (middle and marginal), R-U. Cross section of leaf, X. Basal cells of leaf.

Habitat & Distribution

It is found in a wide variety of habitats such as on rotten logs, bark, *etc.* Distributed in China, Thailand, Java, Papua New Guinea, Philippines, Sabah, Oceania, Australia and India (Tamil Nadu, Kerala) [7, 23, 25].

Specimens Examined

India, Kerala, Kannur, Aralam WLS, Paripputhodu (150 m) *MCN, 87609, 87625* (06.02.2003); Idukki, Munnar (1100 m), *MCN 106911* (25.03.2010).

Key to the Species of *Leucophanes*

1a. Costa deeply channelled above; costal leucocysts in 3-4 layers near the leaf base... *L. glaucum*

1b. Costa almost flat above, not deeply channeled; costal leucocysts in 2-3 layers near the leaf base... *L. octoblepharoides*

Leucophanes glaucum (Schwägr.) J. Proc. Linn. Soc., Bot., Suppl. 1: 25. 1859. *Leucophanes albescens* Müll. Hal., Bot. Zeitung (Berlin) 22: 347. 1864. *L. australe* Oefvers. Förh. Finska Vetensk-Soc. 33: 96. 1891. *L. compactum* Broth., Gen. Musc. Frond. 86. 1900. *L. glaucescens* Müll. Hal. ex M. Fleisch., Die Musci Buitenzorg 1: 178. 23. 1904. *L. sordidum* Müll. Hal., Forschungsr. Gazelle 4(5): 50. 1889. *L. subglaucescens* Müll. Hal. ex Dixon, J. Linn. Soc., Botany 43: 296. 1916.

Plants small, 0.5–1.0 cm high, whitish or greyish to bluish green or yellowish green in dense tufts or cushions. Stem simple or branched, loosely foliate below, densely foliate above. Leaves up to 5.0 mm long, narrowly to linear-lanceolate from oblong-ovate sheathing base, upper part keeled or channelled, acute to apiculate at the apex, flexuose when dry; margins distinctly bordered, with 4–6 rows of linear, hyaline cells throughout, serrulate to serrate above mid-leaf, entire below; costa distinctly serrate on the dorsal side in the upper part of the leaf, costal hyalocyst in 3–4 layers near leaf base, chlorocysts in cross section quadrangular. Gemmae commonly produced at the leaf apex. Dioicous. Outer perichaetial leaves not differentiated from vegetative leaves. Seta slender, upto 9 mm long (Fig. **10 A-O**).

Habitat & Distribution

Leucophanes glaucum is corticolous, and also occurs on logs. It has been reported from Indonesia, China, Japan, Thailand, Vietnam, Malaysia, Philippines, Papua New Guinea, Australia, Thailand and the Nicobar Islands as *L. albescens*. In India it is distributed in Kerala and Tamil Nadu [7, 18].

Specimens Examined

Kerala, Kannur, Theyyottukavu (50 m) *KPR 84687*, Aralam WLS, Chavachi (180 m), *MCN 87583*; Kozhikode, Kakkayam, Ambalappara (1050 m) *MCN 120214*,

Vattakkayam (1000 m) *MCN 120184*; Idukki, Munnar (1100 m) *MCN 106810* (CALI).

Leucophanes octoblepharoides Brid., Bryol. Univ. 1: 763. 1826; Gangulee, Moss. E. India 1(2): 436. 1972. *Syrrhopodon octoblepharis* Nees & Schwaegr., Sp. Musc. Suppl. 4: 311. 1842. *Octoblepharum octoblepharoides* (Brid.) Mitt., Rep. Sc. Res. Voyage Challenges 1(4): 259. 1885.

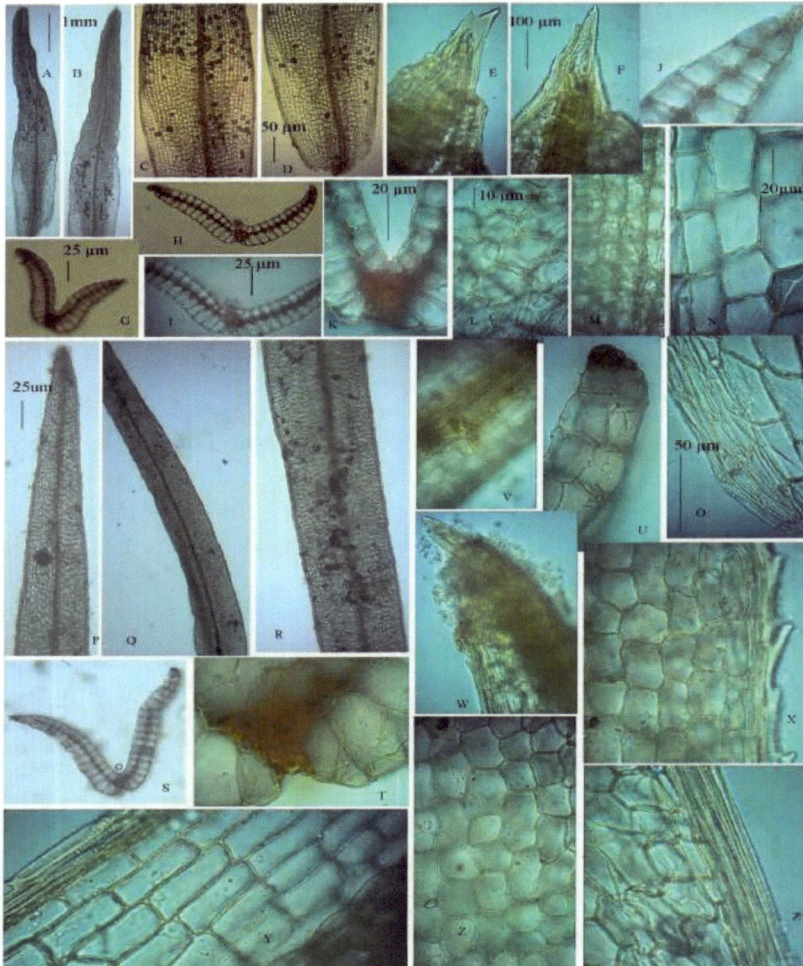

Fig. (10). A-O. *Leucophanes glaucum*: A-B. Leaves, C-D. Basal portion of leaf, E-F. Apical portions of leaf, G-I. Cross section of leaves, J. Cross section of lamina tip, K. Cross section of lamina at midrib region, L-M. Apical cells of leaf, N-O. basal cells of leaf; P-Z1: *Leucophanes octoblepharoides*: P-R. Leaf portions, S. Cross section of leaf, T. Cross section at median region, U. Cross section of lamina portion, V. Costa at tip region, W. Leaf tip, X. Apical marginal portion of leaf, Y-Z. Basal cells of leaf, Z1. Median cells of leaf.

Plants pale to whitish-green, glossy, densely covered with leaves, 2-4 cm high, erect to drooping, usually simple, without central strand; leaves rather crowded, erect or widely spreading, narrowly lanceolate, slightly concave at base, flat above, 1-3 mm long; costa occupying entire upper leaf, but narrowing into leaf base slightly toothed abaxially near the apex, a narrow strand of stereids running close to or at the dorsal surface extending the entire leaf length, hyalocysts occur in the lower leaf base, in three or more layers; lamina cells confined to the marginal part of the leaf base, hyaline, shortly rectangular, 20-25 µm wide, leaves bordered all around with 2-3 rows of very long narrow yellowish cells; sporophyte not common (Fig. **10 P-Z1**).

Habitat & Distribution

On rocks and on logs. It is an Indo-Pacific species distributed in South India (Kerala, Tamil Nadu), North-east India (Assam, Bengal), Australia, Sri Lanka, Bangladesh, Myanmar (Burma), Java, New Guinea, Nepal, Philippines, Sumatra, Tahiti, Admiral Island and Pacific Islands [7, 18].

Specimens Examined

Wayanad, Kalpetta (400 m) *MCN 99665a*; Thiruvananthapuram, Athirumala (1300 m) *MCN 106328,* 12.12.2002 *(CALI).*

Octoblepharum albidum : Hedw., Sp. Musc. 50. 1801; Gangulee, Moss. E. India 1(2): 441. 1972; Mohamed *et al.*, J. Bombay Nat. Hist. Soc. 83: 689. 1986; M. C.Nair & Madhus., J. Econ. Taxon. Bot. 25: 573. 2001. *Octoblepharum minus* Hamp., Vid. Medd. Naturh. For. Kjobenh, ser. 4, 1: 83. 1879. *O. cuspidatum* C.Muell., Gen. Musc. Fr. 88. 1900. *O. martinicense* Mitt. in Cardot, Mem. Soc. Sc. Nat. Cherbourg 32: 41. 1901, *nom. nud. O. pallidum* Besch., *Ibid.* 32: 41. 1901, *nom. nud. O. ekmanii* Ther., Mem. Soc. Cubana Hist. Nat. 13: 220. 1939.

Plants whitish-green, up to 3 cm long; stem without central strand; leaves crowded, erect spreading on a very short stem; leaves linear-lingulate, from an oblong or narrowly obovate, concave base; same in dry and normal (moist) condition, 3 mm to 1 cm long, sheathing base broad, apiculate at tip, minutely serrate; costa wide, smooth on back, with a median row of triangular chlorocyst cells between 5-6 layers of leucocysts in the middle of the leaf and only 2 such layers on the sides; leaf base flanked by 5-8 rows of hyaline linear cells, outer two rows with narrow linear cells, inner rows are rectangular similar to leucocyst cells; seta straight, 5 mm long; capsule erect, oblong-ovoid, peristome 8 teethed, yellowish, operculum conical; spores finely papillose, light brownish, 18-20 µm in diameter (Fig. **9 L-X**).

Habitat & Distribution

It is a widely distributed species seen in a variety of habitats such as on exposed roots, bark, branches, logs, on soil and rocks in scattered colonies, from lower to medium altitude (up to 900 m) mostly in all type of vegetation. It is a common species, reported from South India (Kerala, Karnataka, Tamil Nadu) North-east India (Kumaon, Sikkim), Sri Lanka, Java, New Guinea, China, Bolivia, Columbia, Madagascar, Myanmar, Nepal, Peru, Philippines and Venezuela [7, 18].

Specimens Examined

India, Kerala, Wayanad, Kurichiad range, Dhottakulachi (839 m) *MCN 84552*, Muthanga range, Mankolli (893 m) *MCN 84521a*, Mananthavady (700 m) *MCN, PVM & KPR 80113a;* Kottayam, Vazhur (100m) *MCN* 84638; Palakkad, Parambikulam WLS (750m) *MCN 106811;* Kozhikode, Kakkavayal (100 m) *Reshma, MCN & KPR 1167*; Kakkodi (sealevel), *MCN* 106362; Kannur, Aralam WLS, *Chavachi* (150m), *MCN 87577* (ZGC).

Key to the Species of *Syrrhopodon*

1a. Cancellinae extending to above midleaf .... *S. rufescens*
1b. Cancellinae confined to leaf base 2
2a. Leaves lingulate triangular or elongate triangular often with filamentous gemmae produced just above the hyaline base from lateral cells in the ventral surface of the costa... 3
2b. Leaves narrowly lanceolate or linear, gemmae not produced as described above 4
3a. Cells of chlorophyllose lamina acutely protuberant from the ventral leaf surface, dorsally unipapillose, apex of hyaline lamina acute, leaf margin serrated almost from base to apex, teeth sharp .... *S. parasiticus*
3b. Cells of chlorophyllose lamina acutely protuberant from the ventral leaf surface, often papillose dorsally and ventrally, leaf margin entire, apex minutely serrate .... *S. gardneri*
4a. Leaves gradually narrowing at apex of hyaline lamina margin .... *S. prolifer*
4b. Leaves abruptly narrowing from subrectangular elliptical hyaline leaf base into a linear chlorophyllose limb, margin spinose at shoulders, regularly toothed in distal chlorophyllose limb, from base of chlorophyllose limb to above mid – limb entire .... *S. spiculosus*

Syrrhopodon gardneri (Hook.) Schwaegr. in Hedw., Sp. Musc. Suppl. 2(1): 110. 131. 1824. *Calymperes gardneri* Hook., Musc. Exot. 2: 146. 1819. *C. hobsonii*

Grev., Ann. Lyc. Nat. Hist. New York 1(2): 271. 23. 1825. *C. welwitschii* Duby, Mem. de la Soc. de Phys. et d'Hist. Natur. de Geneva 21: 444. 4 f.8. 1872. *Syrrhopodon strictus* Thwaites & Mitt., J. Linn. Soc. Bot. 13: 299. 1873. *S. aculeatoserratus* Besch., Annal. des Sci. Natur. Botanique ser. 6, 9: 349. 1880. *S. welwitschii* (Duby) Besch., Annal. des Sci. Natur. Botanique, ser. 8, 1: 307. 1896. *S. curranii* Broth., Philippine J. Sci. 5: 142. 1910.

Plants yellowish-green, erect to flexuose, up to 5 cm long and 1 mm wide; leaves narrowly to broadly linear–lanceolate from an erect elliptical hyaline base, ending in a pointed denticulate apex, base sheathing, 6-8 mm long and 0.8 mm wide, margin serrated in distal hyaline leaf base, distally formed by a polystratose rib, which at intervals gives rise to acute (often geminated) teeth; costa ending just short of leaf apex in a dentate tip; hyaline basal lamina sharply defined, marginal ribs in proximal leaf base entire to dentate, composed of long narrow thick walled cells, from around mid-hyaline base to above shoulders rib lacking, margin denticulate, above shoulders composed of quadrate to shortly rectangular cells, 50-58 x 22-26 µm, cells of chlorophyllose lamina often broadly incurved to below leaf apex, cells quadrate to shortly rectangular, in distal leaf mostly 10-12 x 7-10 µm, projecting acutely from the ventral leaf surface, each projection bearing one or more small papillae dorsally and ventrally, dorsally uni-or pluripapillose; gemmae sometimes produced from the ventral surface of the costal apex (Fig. **8 G-N**).

Habitat & Distribution

On tree trunks in semi-evergreen forests. *S. gardneri* (Hook.) Schwaegr. is a pantropical species. In India it is recorded from the North-western Himalayas, West Bengal, Khasi hills, Tamil Nadu, Kerala) [7, 12, 18].

Specimen Examined

Kerala, Wayanad, Chembra hills (1720 m) *MCN 99680 (CALI)*.

Syrrhopodon prolifer Schwägr., Spec. Musc. Suppl. 2(2): 99. 1827.

Plants in loose tufts, pale green, very short stem, about 1-1.5 cm tall, stem simple; leaves monomorphic, twisted and contorted when dry, 1.6-2 mm long, linear to broadly linear, not tapering except at apex, boarded all around with elongate stereidial cells, rhizoids at the base, reddish-purple, cells short rectangular, not bulging, 5x7 µm, papillose laminal cells, very low, rough, rarely smooth, present on dorsal and ventral surface (Fig. **11 A-E**).

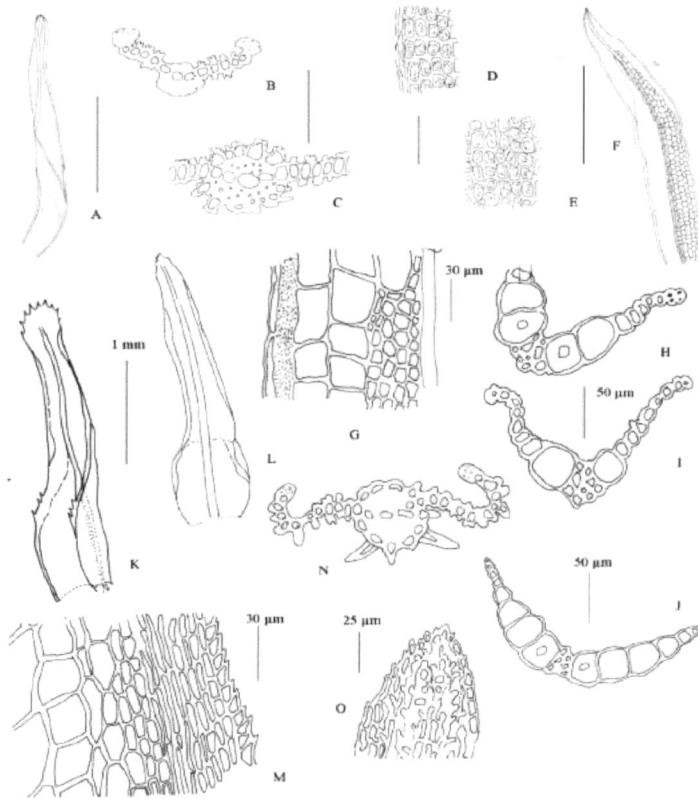

Fig. (11). A-E. *Syrrhopodon prolifer*: A. Leaf, B-C. Section through chlorophyllose limb, D. Margin of chorophyllose lamina, E. Chlorophyllose cells; F-J: *Syrrhopodon rufescens*, F. Leaf (dorsal view), G. Cells at chlorophyllose lamina (hyaline cells end), H-I. Leaf section (median), J. Leaf section (basal); K-O: *Syrrhopodon spiculosus*, K-L. Leaves, M. Marginal-basal cells, N. Section through chlorophyllose cells, O. Leaf apex.

Habitat & Distribution

It occurs in a wide range of habitat; present collection is from rocky patch. Known from India (Kerala, Tamil Nadu), Mexico, Panama, Costa Rica, Venezuela, Bolivia, Surinam and Brazil [7, 12]. The present collection is a new record for Kerala.

Specimen Examined

Kerala, Idukki, Eravikulam National Park (1400 m) *MCN 80175* (CALI).

Syrrhopodon parasiticus (Sw. ex Brid.) Besch. in Ann. Sci. Nat. Bot. ser. 8, 1: 298 (1895). *Bryum parasiticum* Brid., Muscol. Recent. 2(3): 54 (1803).

Syrrhopodon wiemansii M.Fleisch., Musc. Fl. Buitenzorg 1: 204, 210 (1904).
Calymperopsis wiemansii (M.Fleisch.) M.Fleisch. in Biblioth. Bot. 80: 5 (1913).
Calymperopsis parasitica (Brid.) Broth., Nat. Pfanzenfam. 2nd ed., 10: 235 (1924).

Plants 0.5-1.5 cm long, light to dark greenish, densely matted with rhizoids below, highly branched, plants contorted when dry; leaves more crowded towards the tip, 3.5-4 mm long with oblong hyaline base narrowing slightly into a lingulate to lanceolate chlorophyllose limb, bordered from base to beyond mid-leaf by 2-4 rows of hyaline elongated cells (stereids), distal margin minutely serrate, leaf apex broadly acute. Costa prominent, ending in leaf apex, smooth, ventral surface from just above hyaline base to midleaf composed of shortly quadrate to rectangular cells; chlorophyllose lamina incurved, cells of chlorophyllose lamina acutely protuberant, dorsally unipapillose, cells isodiametric to slightly longer than broad; hyaline leaf base sharply defined, with a long acute apex penetrating the chlorophyllose lamina, cells large rectangular, leaf margin denticulate from base to apex, teeth sharp, above hyaline base consisting of a thin, intermittent to continuous strand of steriods, in hyaline base consisting of a broad rib of linear thin to thick walled cells, merging distally into chlorophyllose lamina; gemmae long, filamentous, uniseriate, produced from either side of ventral surface of the costa (Fig. **12 A-P**).

Habitat & Distribution

Epiphytic on branches and twigs. Pantropical species, known in India only from Kerala [18].

Specimens Examined

India, Kerala, Wayanad, Kuruva Island, 800m, 25.09.2013 *Manjula 861b, 865b & 862b (ZGC)*.

Syrrhopodon rufescens Hook. & Grev., Edinburgh J. Sci. 3: 227 (1826).
Leucophanella rufescens: (Hook. & Grev.) M. Fleisch., Musc. Fl. Buitenzorg 1: 200. 1904.

Plants 1-3 cm high, densely covered with leaves forming green cushions, rhizoids yellowish to reddish; leaves 1-2 mm long, suberect to patent recurved, lanceolate, apex acute, largely composed of hyaline lamina (Cancellinae); costa thin ending at apex, composed of stereids with a median layer of guide cells, two cells wide, superficial cells differentiated, subrectangular, many giving rise to teeth, towards leaf base elongated and smooth; chlorophyllose lamina very short, narrowly tapering down either side of hyaline lamina, cells quadrate to subrectangular, irregularly rounded-elliptical, 7-12 x 6-10μm, apical papilla present; hyaline lamina sharply defined, cells large thin walled, 5-6 rows on either side of the costa; leaf margin entire from base to apex, formed by a thin strand of stereids/substereids (Fig. **11 F-J**).

Fig. (12). *Syrrhopodon parasiticus*: A. Habit, B. Branch (enlarged), C. Apical cells of leaf, E. Median marginal cells of leaf, F. Calcinnae joining the cholorophyllous cells, G. Leaf base, H. Basal cells of leaf, i. Basal cells of leaf (enlarged), J. Marginal cells of leaf, K. Chlorophyllous cells at middle portion, L-M. Cross section of the leaf tip, N. Cross section of leaf (median part), O-P. Cross section at base (figs, C,F,G same size; D, E, H, P same size).

Habitat & Distribution

On tree trunks. Distributed in India and Malaysia, Java, Philippines [7, 26].

Specimen Examined

Kerala, Ernakulam, Thattekkadu Bird Sanctuary, *Nikesh 12013* (CALI).

Syrrhopodon spiculosus Hook. & Grev. In Edinburgh J. Sci. 3: 226. 1825.

Plants 1-3 cm high, leaves 3-4 mm long with an erect subrectangular-elliptical semisheathing hyaline base, distally narrowing into a linear chlorophyllose limb with dentate blunt apex; costa ending just below apex, above hyaline leaf base on dorsal and ventral surfaces giving rise to acute spines; cells of chlorophyllose lamina isodiametric or slightly longer than broad, rounded to elliptical with 4-6 sides, cells possess coronate-papillose projections, which towards the leaf apex become more spine-like; hyaline lamina sharply distinct; leaf margins at apex dentate, above shoulders of leaf to a short distance below apex often incurved, consisting of an entire polystratose rib of stereids, at shoulders giving rise to a row of long acute spines, below shoulders entire (Fig. **11 K-O**).

Habitat & Distribution

On logs and tree trunks in high altitude areas. A paleotropical species reported from Philippines, India (Kerala). *S. spiculosus* Hook. & Grev. has been collected by Daniels *et al.* [27] from Silent Valley National Park in Palakkad district along with *Indopottia zanderi* Daniels, Raja & Daniels. However, for the present report we could not locate this species in Silent Valley NP, but collected it from New Amarambalam Reserve Forest of Nilambur area.

Specimens Examined

Kerala, Malappuram Dt., New Amarambalam Reserve Forest (1200 m) *KPR 27249* (CALI).

CONSENT FOR PUBLICATION

Not applicable.

CONFLICT OF INTEREST

The authors confirm that this chapter contents have no conflict of interest.

ACKNOWLEDGEMENTS

The authors are indebted to Dr. L.T. Ellis, Natural History Museum, London, and Dr. Noris Salazar Allen, Smithsonian Tropical Research Institute, United States of America for confirming the identity of most of our collections. We are also grateful to the staff members of the Kerala Forest Department for extending support during our field studies. We thank Kerala State Council for Science Technology & Environment (KSCSTE), Thiruvananthapuram for financial support. We are also thankful to the authorities of the Zamorin's Guruvayurappan College, Kozhikode for giving the support.

REFERENCES

[1]   Edwards SR. A revision of west tropical African Calymperaceae I. Introduction and Calymperes. J Bryol 1980; 11: 49-93.
[http://dx.doi.org/10.1179/jbr.1980.11.1.49]

[2]   Gradstein SR, Pocs T. Bryophytes In: Lieth H, Werger MJA, Eds. Tropical rain forest ecosystems. Amsterdam: Elsevier Sci. Publ. B.V. 1989.
[http://dx.doi.org/10.1016/B978-0-444-42755-7.50022-5]

[3]   Reese WD. Calymperaceae. Flora Neotropica 1993; 58: 1-102.

[4]   Kindberg NC. Genera of European and North American Bryineae (Mosses)/synoptically disposed by Kindberg NC, Linkoeping. Sweden: P.M. Sahlstroems Bookselling (C.V. Zickerman). 1897.

[5]   Andrews AL. Taxonomic notes VI. The Leucobryaceae. Bryologist 1947; 50: 319-29.
[http://dx.doi.org/10.1639/0007-2745(1947)50[319:TNVTL]2.0.CO;2]

[6]   Buck WR, Goffinet B. Morphology and classification of mosses. In: Shaw A-J, Goffinet B, Eds. Bryophyte Biology. Cambridge: Cambridge University Press 2000; pp. 71-123.
[http://dx.doi.org/10.1017/CBO9781139171304.004]

[7]   Gangulee HC. Mosses of Eastern India and adjacent regions. Calcutta: Fasc. 2&3. 1971; 1.

[8]   O'Shea BJ. Checklist of the mosses of sub-Saharan Africa (version 5, 12/06). Tropical Bryology Research Report 2006; 6: 1-252.

[9]   Reese WD. *Calymperes* (Musci: Calymperaceae): World ranges, implications for patterns of historical dispersion and speciation, and comments on phylogeny. Brittonia 1987; 39: 225-37.
[http://dx.doi.org/10.2307/2807380]

[10]   Ellis LT. Taxonomic notes on *Calymperes*. J Bryol 1987; 14: 681-90.
[http://dx.doi.org/10.1179/jbr.1987.14.4.681]

[11]   Ellis LT. Taxonomic notes on *Calymperes* II. J Bryol 1988; 15: 127-40.
[http://dx.doi.org/10.1179/jbr.1988.15.1.127]

[12]   Ellis LT. Taxonomic revision of *Calymperes* in Southern India and neighbouring Islands. J Bryol 1989; 15: 697-732.
[http://dx.doi.org/10.1179/jbr.1989.15.4.697]

[13]   Reese WD. Nomenclature of palaeotropical Calymperaceae, with description of *Syrrhopodon meijeri* sp. nov. Bryologist 1988 ['1987']; 90201.

[14]   Ellis LT, Tan BC. The moss family Calymperaceae (Musci) in the Philippines. Bull Nat Hist Mus. London (Botany) 1999; 29(1): 1-46.

[15]   Ellis LT, Eddy A. Leucobryaceae to Buxbaumiaceae 2. London: Handb Males Mosses Natural History Museum Publications 1990. iv.

[16]   Manju CN, Rajesh KP, Madhusoodanan PV. Checklist of the Bryophytes of Kerala, India. Tropical Bryology Research Report 2008; 71.

[17]   Salazar N, Tan BC. *Octoblepharum arthrocormoides* (Calymperaceae) N. Salazar Allen & B.C. Tan, sp. nov., a new species from Tropical Asia. Botany 2010; 88(4): 440.

[18]   Nair MC, Rajesh KP, Madhusoodanan PV. Bryophytes of Wayanad in Western Ghats. Malabar Natural History Society, Kozhikode. 2005. i-iv + 284pp.

[19] Manju CN, Manjula KM, Rajesh KP. *Syrrhopodon parasiticus* (Calymperaceae: Bryophyta) a new record for India. In: New National and regional bryophyte records (Ellis *et al.* 2015). J Bryol 2015; 37(1): 68-85.

[20] Reese WD. Synopsis of *Syrrhopodon* subgenus *Pseudocalymperes*. Bryologist 1995; 98: 141-5. [http://dx.doi.org/10.2307/3243650]

[21] Tixier P. A contribution to the knowledge of the mountain moss flora of Sri Lanka. Ceylon Journal of Science, n ser (Biol Sci) 1975; 11(2)123.

[22] Rajeevan B. Studies on the Bryophye flora of the Idukki District, Kerala 1990.

[23] Manju CN, Rajesh KP, Jitha S, Reshma PK, Prakashkumar R. Bryophyte diversity of Kakkavayal Reserve Forest in the Western Ghats, Kerala. Arch Bryol 2011; 108: 1-7.

[24] Eddy A. A handbook of Malesian mosses. London 1990; Vol. 2.

[25] Ellis LT. A taxonomic revision of *Exodictyon* Card. (Musci: Calymperaceae). Lindbergia 1985; 11: 9-37.

[26] Bartram EB. Supplement to the manual of Hawaiian mosses. Occas Pap Bernice P Bishop Mus 1939; 15: 93-108.

[27] Daniels AED, Raja RDA, Daniel P. *Indopottia zanderi* (Bryophyta, Pottiaceae) gen. et sp. nov. from the Western Ghats of India. J Bryol 2010; 32: 216-9. [http://dx.doi.org/10.1179/037366810X12578498136390]

CHAPTER 6

Bryodiversity of Wilson Hill and Surrounding Area of Valsad, Gujarat

Rakesh V. Gujar and **Dharmendra G. Shah**[*]

Ecotoxicology & Lower Plants Lab., Department of Botany, The Maharaja Sayajirao University of Baroda, Vadodara 390 002, Gujarat, India

Abstract: Valsad district lies in the southern part of the state which harbours some locations embracing good diversity of bryophytes. In the present study, the bryodiversity of a potential location, *i.e.*, Wilson hill and surroundings has been explored and reported. The study reveals the existence of 15 species of bryophytes that included 5 liverworts, 2 hornworts and 8 mosses under 9 families. Three new additions to the Bryoflora of Gujarat have also reported, *viz.*, *Taxilejeunea ghatensis, Notothylas anaporta,* and *Notothylas himalayensis.* Interestingly, the majority of species found were terricolous in nature.

Keywords: Mosses, Liverworts, Hornworts, South Gujarat.

INTRODUCTION

Gujarat is the sixth largest state in India and lies in the Western part of India. It covers an area of approximately 1,96,022 km^2 [1] with 18,827 km^2 [2] forest area. The study of the lower groups, especially, bryophytes have been neglected in the state. The major work consists of "Bryophyte Flora of Gujarat" [3]. It includes 18 locations across Gujarat and several districts, including Vadodara, Narmada, Bharuch and Surat that were not covered in the flora. In the south Gujarat, only 6 sites were visited. Moreover, Valsad District was not included in the study (Fig. **1**). There are few locations that were potentially bryo-rich but not covered. One of these locations is Wilson hill. It is named as 'Wilson hill' in the honour of Lord Wilson – the governor of Mumbai during 1923-1928. It has a good forest track with some scenic locations in the Pangarbari Wildlife Sanctuary. This location was not at all visited by past workers and no reports related to bryophytes are found from the district. This work is the first kind of data gathered form the place.

[*] **Correspondence author Dharmendra G. Shah:** Ecotoxicology & Lower Plants Lab., Department of Botany, The Maharaja Sayajirao University of Baroda, Vadodara 390 002, Gujarat, India; Tel: +91-9427336842; Email: shahdhamu@gmail.com

Afroz Alam (Ed.)

MATERIALS AND METHOD

Study Area

Gujarat lies on the western side of the India. It has been divided into 33 districts. It mainly has an arid or semiarid region like Kutch and Saurashtra and South Gujarat that receives good rainfall *viz.*, Dangs, Valsad, Navsari, Tapi Surat *etc.* The District Valsad is unique in its atmosphere with Wilson Hill, a hill station situated at a distance of 60 km from Valsad town. It comes in type 3B Tropical moist deciduous forest [4]. Its average elevation is 750 m above the sea level. During the winter, the weather is cool and in summer, it becomes warm with occasional cool breezes. The average rainfall throughout the year is about 2000 mm. Near the Wilson hill, small waterfall locally called as *Shankar Dhodh* also exists with 30 m high waterfall.

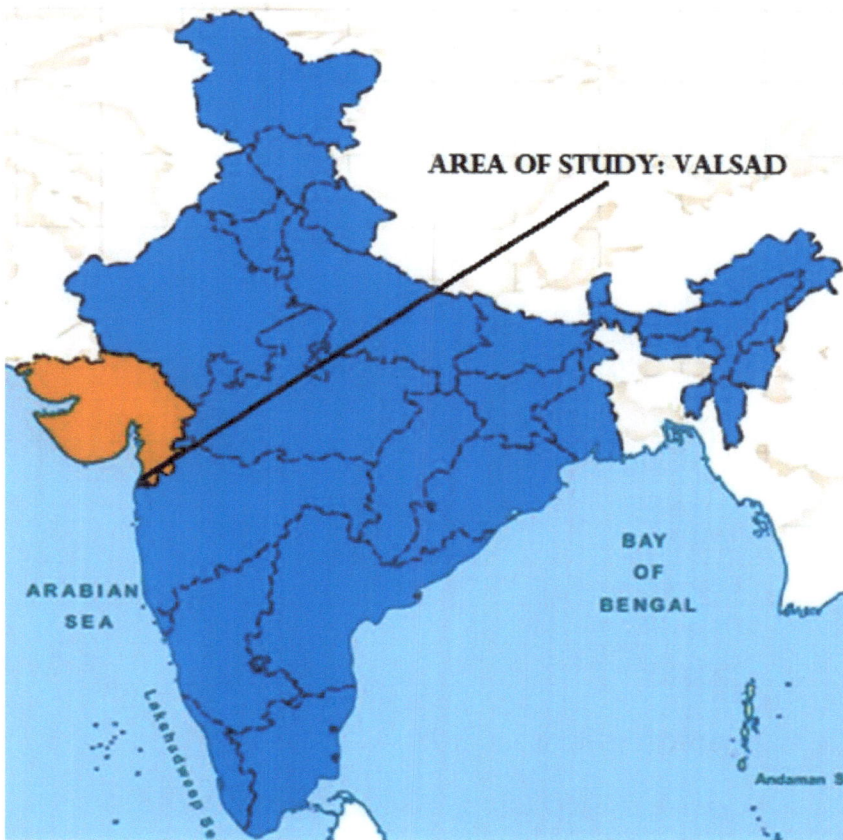

Fig. (1). Map of India showing Valsad (Gujarat).

Sample Collection and Analysis

The samples were collected from different locations from Wilson hills and surrounding. The samples were collected in paper bags with the help of a scalpel or knife. The samples were preferably collected in sporophytic stages. The samples were collected in brown bags and air dried for voucher preparation. The brown bags also consist of a label with the preliminary information of substratum, temperature, type of sample, GPS location using a Garmin E-trex10 GPS receiver *etc.* Photography of the relevant parts was taken in the field using Canon Powershot A480. The samples are identified with the help of standard available literature [5 - 11]. The confirmation of some samples was done from National Botanical Research Institute (NBRI) Lucknow. The Voucher specimens were deposited in the BARO herbarium. The authentication of the scientific name and its present status was checked from the "the Plant List".

Table 1. Diversity of bryophytes in Wilson hill and surrounding with Habitat.

Sr. No	Name of Species	Family	Habitat
1	*Rectolejeunea aloba* (Sande Lac.) Steph.	Lejeuneaceae	Bark
2	*Taxilujenea ghatensis* Verma & S.C. Srivast.	Lejeuneaceae	Bark
3	*Cyathodium cavernarum* Kunze	Targioniaceae	Soil
4	*Riccia bilardieri* Mont. *et* Nees.	Ricciaceae	Soil
5	*Asterella angusta* (Steph.) Kachroo.	Ayotoniaceae	Wall
6	*Notothylas anaporta* Udar *et* Singh.	Notothylaceae	Soil
7	*Notothylas himalayensis* Udar *et* Singh.	Notothylaceae	Soil
8	*Fissidens zollingeri* Mont.	Fissidentaceae	Bark
9	*Hyophila involuta* (Hook.) Jaeg.	Pottiaceae	Rock
10	*Hydrogonium consanguineum* (Thw. *et* Mitt.)	Pottiaceae	Soil
11	*Gymnostomiella vernicosa* (Hook.) Fleisch.	Splachnaceae	Wall
12	*Anomobryum auratum* (Mitt.) Jaeg.	Bryaceae	Soil
13	*Brachymenium turgidum* Broth. *ex* Dix.	Bryaceae	Bark
14	*Bryum coronatum* Schwaegr.	Bryaceae	Wall
15	*Bryum capillare* (L.) Hedw.	Bryaceae	Soil

RESULTS AND DISCUSSION

A total 20 samples were collected from which the identity of 15 samples could be established, during the study. 15 species were recorded from 9 families containing 5 liverworts, 2 hornworts and 8 mosses (Table **1**). Among liverworts 5 genera, mosses 7 genera whereas in hornwort only one genera are reported. *Taxilejeunea ghatensis, Notothylas anaporta* and *Notothylas himalayensis* are new to Gujarat state. Valsad lies close to the Dangs district that hold a small portion of Western Ghats hence an endemic species of the Western Ghat, the leafy liverworts - *Taxilejeunea ghatensis* was discovered from the area. Before this study, there were no reports of leafy liverworts from the area, but now there are two leafy liverworts in the state. These species occupy different habitats (Fig. **2**), some occur on soil (7 sp.), some on bark (4 sp.), wall (3 sp.) and on rock (1 sp.). Thus, terricolous species were more than phycocolous and corticolous. The second most dominant (27%) habitat was of bark (epiphyte) as the elevation and other topographic parameters favours growth of many epiphytic liverworts and mosses since the precipitation and humidity is higher than the other parts of the district. In mosses, Bryaceae appeared as the dominant family, while in case of hornworts, Notothylaceae was reported with only one genus. Most of the habitats suitable for the growth of bryophytes had been found in the area (Fig. **3**). The pocket of Wilson hills in many places has grasslands that kept the surface moist and thus favoured the diversity of terricolous bryophytes.

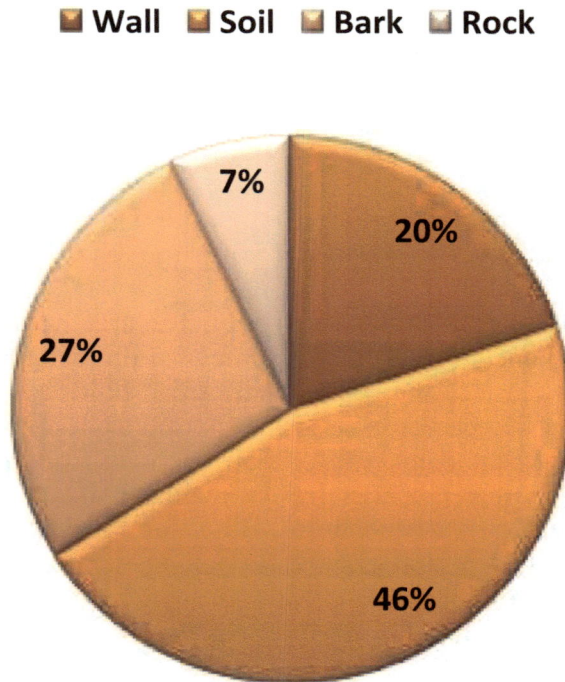

Fig. (2). Percentage distribution of bryophyte's habitat.

Fig. (3). Bryorich habitats of the study area.

CONCLUSION

The study area was rich in bryophyte diversity. The area lies near the Pangarbari Wildlife Sanctuary, hence, diversity was somewhat protected and conserved to a larger extent as it is close to the Dangs district that holds a small portion of the Western Ghats. Total 77 species have been reported from all over the Gujarat till date. Among them in the present study, 15 are reported from this small pocket of the state. This shows that the area has congenial habitats, thus favoured the growth of diverse terricolous bryophytes. Based on this study, it is evident that Gujarat should be explored more for the bryophytes before they become extinct and remain highly unexplored.

CONSENT FOR PUBLICATION

Not applicable.

CONFLICT OF INTEREST

The authors confirm that this chapter contents have no conflict of interest.

ACKNOWLEDGEMENTS

Authors are greatly thankful to the Gujarat Biodiversity Board, Gandhinagar, for their financial support. The authors are also thankful to Dr. A. K. Asthana and Dr. Vinay Sahu for helping in the identification and confirmation of the plant specimens.

REFERENCES

[1]   Anon. Census of India http://www.censusindia.gov.in (Accessed: 2nd March 2016), 2013.

[2]   Anon. State of Forest Report, 2011, Forest Survey of India. Dehradun, India: Ministry of Environment and Forests 2011.

[3]   Chaudhary BL, Sharma TP, Charu S. Bryophyte Flora of Gujarat, India. Udaipur, India: Himanshu Publications 2006.

[4]   Champion HG. A Preliminary survey of the Forest types of India and Burma. Indian Forester 1937; 1: 1-286.

[5]   Bapna KR, Kachroo P. Hepaticology in India & II. Udaipur, India: Himanshu Publications 2000.

[6]   Chaudhary BL, Deora GS. Moss Flora of Rajasthan (India). Udaipur, India: Himanshu Publications 1993.

[7]   Chaudhary BL, Sharma TP, Bhagora FS. Bryophyte Flora of North Konkan Maharashtra, India. Udaipur, India: Himanshu Publication 2008.

[8]   Gangulee HC. Mosses of Eastern India and Adjacent regions. 1969–1980.

[9]   Singh DK. Notothylaceae of India and Nepal - A Morphotaxonomic revision. Dehradun, India: Bishen Singh Mahendra Pal Singh 2002.

[10]   Aziz MN, Vohra JN. Potticeae (Musci) of India. Dehradun, India: Bishen Singh Mahendra Pal Singh 2008.

[11]   Alam A, Pandey S, Singh V, Sharma SC, Sharma V. Moss flora of Mount Abu (Rajasthan), India: An updated checklist. Tropical Plant Research 2014; 1(1): 8-13.

CHAPTER 7

Analysis of Microsatellites in Mitochondrial Genome of *Aneura pinguis* (L.) Dumort

Sonu Kumar and **Asheesh Shanker**[*]

Department of Bioinformatics, Central University of South Bihar, Gaya- 824236, India

Abstract: Microsatellites also called as simple sequence repeats (SSRs) are stretch of repetitive DNA consisting of 1-6 nucleotides. Microsatellites are widely used as molecular markers and have been identified in many organisms including Bryophytes. However, despite the availability of the mitochondrial genome of *Aneura pinguis,* the information about its mitochondrial SSRs (mtSSRs) is not well understood. In the present study, a total of 26 mtSSRs were mined in the mitochondrial genome of *Aneura pinguis*. Di-nucleotides (15, 57.69%) were the most abundant followed by tetra-nucleotides (6, 23.08%), tri-nucleotides (4, 15.38%), and mono-nucleotide (1, 3.85%) repeats, whereas penta- and hexa-nucleotides were completely absent. The identified mtSSRs can be used in transferability studies and also play an important role in genetic diversity analysis of *Aneura* species.

Keywords: *Aneura pinguis*, Bryophytes, Microsatellites, Mitochondria.

INTRODUCTION

Bryophytes are small, non-vascular, simplest land plants and are broadly categorized into hornworts, liverworts, and mosses. *Aneura pinguis* (L.) Dumort is a liverwort, which belongs to the family Aneuraceae. *A. pinguis* is a thalloid liverwort with a very simple morphological structure. It is found in diverse geographical regions and grows in various habitats [1, 2]. Previously, the paraphyletic origin of bryophytes has been proposed using chloroplast and mitochondrial genome sequences [3 - 5]. Moreover, the organellar genome sequences of bryophytes have also been used for the identification of microsatellites [6, 7].

Microsatellites also referred as simple sequence repeats (SSRs), are short stretch of repetitive nucleotide sequence and consist of 1-6 nucleotides. The microsatellites are found in both prokaryotic and eukaryotic genomes [8, 9] and also in organellar genomes [10 - 12].

[*] **Corresponding author Asheesh Shanker:** Department of Bioinformatics, Central University of South Bihar, Gaya-824236, India; Tel: +91-9414478655; E-mail: ashomics@gmail.com

Afroz Alam (Ed.)

Microsatellites are widely applied molecular markers due to their codominant and highly reproducible nature. Earlier, microsatellites have been identified in mitochondrial [6, 13] and chloroplast [14 - 16] genome sequences of bryophytes. Apart from this, microsatellites were also identified in various plants, including *Arabidopsis* [17], *Citrus sinensis* [8], and *Cocos nucifera* [18].

Despite the availability of organelle genome sequences in public databases the 748 mitochondrial genome sequences of plants are very few [19] and increasing with a very slow pace. Moreover, the information of mitochondrial SSRs (mtSSRs) in *A. pinguis* is not well understood, therefore, the present study was designed to mine and analyze them.

MATERIALS AND METHODS

Mitochondrial Genome Sequence of *Aneura pinguis*

The mitochondrial genome sequence of *Aneura pinguis* (Accession no. NC_026901.1) was retrieved from the National Center for Biotechnology Information (NCBI; www.ncbi.nlm.nih.gov) in FASTA and GenBank file format.

FASTA is a text-based format that represents nucleotide or protein sequences using their single-letter codes. The first line in a FASTA format started with a ">" (greater-than) symbol used for a unique description and the rest of the line represents the sequence (Fig. **1**). GenBank format contains annotation section, followed by a sequence. The annotation section contains CDS (coding region) along with other information (Fig. **2**).

The length of the complete mitochondrial genome of *A. pinguis* is 165603 bp that contains A (26.2%), C (23.6%), G (23.8%), and T (26.4%) base composition [20].

Mining of Microsatellites

Microsatellites were mined in a retrieved mitochondrial genome sequence of *A. pinguis* with the help of a Microsatellite identification tool (MISA, http://pgrc.ipk-gatersleben.de/misa/misa.html) [21]. The minimum repeat sizes of ≥12 for mono-, ≥6 for di-, ≥4 for tri-, ≥3 for tetra-, penta-, and hexa-nucleotide repeats were considered to mine the microsatellites. Moreover, the distance between the two microsatellites was taken as 0. The identified mitochondrial microsatellites were categorized into coding and non-coding mtSSRs with the help of GenBank file. The coding region is a portion of DNA or RNA that codes for protein, whereas non-coding does not encode protein sequences.

RESULTS AND DISCUSSION

In the present study, a total of 26 perfect microsatellites were mined with a density of 1 microsatellite/6.36 Kb in the mitochondrial genome of *A. pinguis.* The length of the identified microsatellites varied from 12 to 34 nucleotides. Information about identified mitochondrial microsatellites along with their start-end position is presented in Table **1**.

Frequency and Distribution of Microsatellites

Di-nucleotides (15, 57.69%) were the most abundant followed by tetra-nucleotides (6, 23.08%), tri-nucleotides (4, 15.38%), and mono-nucleotide (1, 3.85%) repeats. Penta- and hexa-nucleotides were completely absent in the mitochondrial genome of *A. pinguis* (Fig. **3**).

Among di-nucleotide repeats, motif AT (13, 86.67%) was the most frequent, followed by TA (2, 13.33%). Motif TTAT (2, 33.33%) followed by AGTA, CAAG, CCCT, and CGAA (1, 16.67% of each) was identified among tetra-nucleotides. Among tri-nucleotides, motif AAT, ATT, TAT, and TTC (1, 25% of each) were found. C was the only motif found in mono-nucleotides. The frequencies of identified motifs are presented (Fig. **4**).

Apart from this, the identified microsatellites were also categorized as coding, non-coding, and coding-non-coding. A total of 24 (92.31%) mtSSRs were detected in non-coding and 2 (7.69%) in coding region (Fig. **5**). None of them mtSSRs showed overlap of these regions in the mitochondrial genome of *A. pinguis.*

The total number of identified mtSSRs in *A. pinguis* (26) is lower than *Mielichhoferia elongata* (73) [6]. Earlier, a total of 475 mtSSRs were identified in six bryophyte species (two liverworts *Marchantia polymorpha* and *Pleurozia purpurea*, two mosses *Physcomitrella patens* and *Anomodon rugelii*, and two hornworts *Phaeoceros laevis* and *Nothoceros aenigmaticus*) is also higher than *A. pinguis* [13].

The density of mtSSRs identified in *A. pinguis* (1 SSR/6.36 kb) is higher than density of microsatellite identified in mitochondrial SSRs of *Physcomitrella patens* (1 SSR/2.06 kb) [22] and chloroplast SSRs identified in *Aneura mirabilis* (1 SSR/5.68 kb [23], *Ptilidium pulcherrimum* (1 SSR/5.17 kb [24], *Tetraphis pellucida* (1 SSR/3.04 kb) [7], *Anthoceros formosae* (1 SSR/2.4 kb) [14] and *Marchantia polymorpha* (1 SSR/1.83 kb) [15], whereas lower than *Pellia endiviifolia* (1SSR/7.09 kb) [25]. The difference in the density of identified microsatellites might be due to different parameters including the length of microsatellites taken and the composition of nucleotide sequences mined.

Table 1. Information of mined mtSSRs in *A. pinguis*.

S. No.	Motifs	Size	Start Position	End Position
1.	$(AT)_6$	12	5775	5786
2.	$(AT)_6$	12	9843	9854
3.	$(AT)_{12}$	24	15907	15930
4.	$(AT)_6$	12	16925	16936
5.	$(TAT)_4$	12	26469	26480
6.	$(TTC)_4$	12	31520	31531
7.	$(AT)_6$	12	36041	36052
8.	$(AT)_6$	12	36057	36068
9.	$(C)_{18}$	18	36562	36579
10.	$(AAT)_4$	12	37452	37463
11.	$(AT)_7$	14	41912	41925
12.	$(AT)_6$	12	48783	48794
13.	$(CAAG)_3$	12	61529	61540
14.	$(TTAT)_3$	12	65925	65936
15.	$(CCCT)_3$	12	67121	67132
16.	$(AT)_{13}$	26	68913	68938
17.	$(AT)_{17}$	34	87981	88014
18.	$(TA)_8$	16	88169	88184
19.	$(AT)_7$	14	99279	99292
20.	$(CGAA)_3$	12	112332	112343
21.	$(TTAT)_3$	12	131293	131304
22.	$(TA)_{13}$	26	147156	147181
23.	$(AT)_8$	16	153323	153338
24.	$(ATT)_4$	12	153791	153802
25.	$(AT)_7$	14	155703	155716
26.	$(AGTA)_3$	12	162880	162891

```
>NC_026901.1 Aneura pinguis mitochondrion, complete genome
ACCCCATGAAAATTATCAAAGACCACTTAGAGGCCGCCGGCGGGGGCCGAAACAGCAAAGCCACCCCGCG
CGTCACGAGGCCCGTAGGGCAGTTGACTTTTCCGTGCGTCGCATAAACCGGGAGTGCAGATTCAGGTGCT
                                   .
                                   .
                                   .
TAGAAACAGATTTTGAGTCTGCCGTGTTTGCCGTTCCACCAAGCGAGTCTTGGCTTTTATTAGGAATAGA
TCGAAAATATGCAGCGGTTGCCGCCGCCCCGCCCTGCATTCCTCGGCGACCGC
```

Fig. (1). Representation of FASTA format using mitochondrial genome sequence of *A. pinguis*.

```
LOCUS        NC_026901      165603 bp    DNA      circular PLN 22-APR-2015
DEFINITION   Aneura pinguis mitochondrion, complete genome.
ACCESSION    NC_026901
VERSION      NC_026901.1
DBLINK       BioProject: PRJNA281853
KEYWORDS     RefSeq.
SOURCE       mitochondrion Aneura pinguis
  ORGANISM   Aneura pinguis
             Eukaryota; Viridiplantae; Streptophyta; Embryophyta;
             Marchantiophyta; Jungermanniopsida; Metzgeriidae;
                                     .
                                     .
FEATURES              Location/Qualifiers
     source           1..165603
                      /organism="Aneura pinguis"
                      /organelle="mitochondrion"
                      /mol_type="genomic DNA"
                      /specimen_voucher="Apinguis1"
                      /db_xref="taxon:39026"
                                     .
                                     .
     CDS              join(2240..2965,3584..4840)
                      /gene="nad5"
                      /locus_tag="YB91_gp40"
                      /old_locus_tag="PlpuMp01"
                      /codon_start=1
                      /product="NADH dehydrogenase subunit 5"
                      /protein_id="YP_009132621.1"
                      /db_xref="GeneID:24143138"
                                     .
                                     .
ORIGIN
        1 accccatgaa aattatcaaa gaccacttag aggccgccgg cggggggccga aacagcaaag
                                     .
                                     .
   165541 taggaataga tcgaaaatat gcagcggttg ccgccgcccc gccctgcatt cctcggcgac
   165601 cgc
//
```

Fig. (2). Representation of GenBank format using mitochondrial genome sequence of *A. pinguis*.

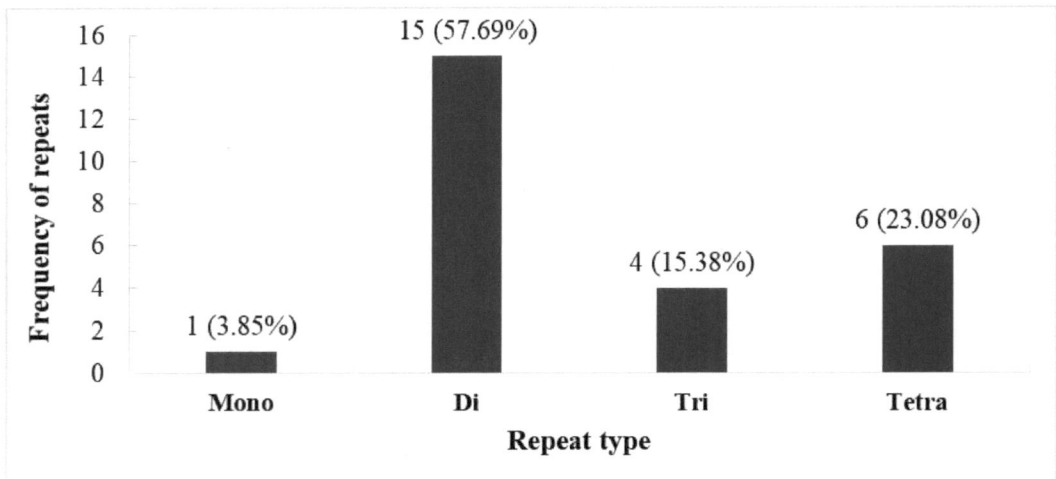

Fig. (3). Frequency of mono - tetra repeats mined in mitochondrial genome of *A. pinguis*.

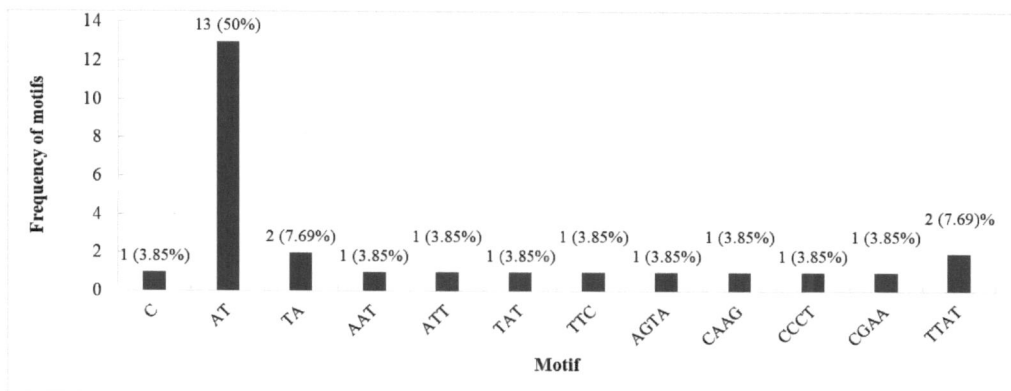

Fig. (4). Microsatellite motifs identified in mined in mitochondrial genome of *A. pinguis*.

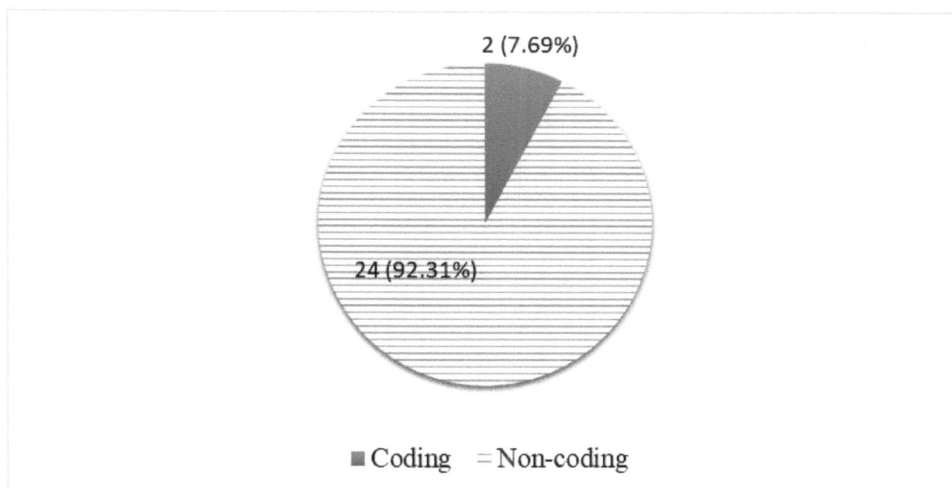

Fig. (5). Distribution of Coding and Non-coding microsatellites detected in mitochondrial genome of *A. pinguis*.

CONCLUSION

The microsatellites were successfully mined in mitochondrial genome sequence of *A. pinguis* and their distribution in coding/non-coding regions was analyzed. The identified mtSSRs will help in transferability analysis and also play a significant role in genetic diversity studies of *Aneura* species.

CONSENT FOR PUBLICATION

Not applicable.

CONFLICT OF INTEREST

The authors confirm that this chapter contents have no conflict of interest.

ACKNOWLEDGEMENT

SK is thankful to the University Grants Commission (UGC), New Delhi, India, to provide financial support in the form of Rajiv Gandhi National Fellowship.

REFERENCES

[1]   Paton JA. The liverwort flora of the British Isles. Colchester: Harley Books 1999.

[2]   Damsholt K. Illustrated Flora of Nordic Liverworts and Hornworts. 1st Ed. Lund: Nordic Bryological Society 2002; pp. 654-6.

[3]   Shanker A. Paraphyly of bryophytes inferred using chloroplast sequences. Arch Bryol 2013; 163: 1-5.

[4]   Shanker A. Inference of bryophytes paraphyly using mitochondrial genomes. Arch Bryol 2013; 165: 1-5.

[5]   Shanker A. Combined data from chloroplast and mitochondrial genome sequences showed paraphyly of bryophytes. Arch Bryol 2013; 171: 1-9.

[6]   Goruynov DV, Goryunova SV, Kuznetsova OI, *et al.* Complete mitochondrial genome sequence of the "copper moss" *Mielichhoferia elongata* reveals independent nad7 gene functionality loss. PeerJ 2018; 6e4350.
        [http://dx.doi.org/10.7717/peerj.4350]

[7]   Shanker A. Detection of simple sequence repeats in the chloroplast genome of *Tetraphis pellucida* Hedw. Plant Sci Today 2016; 3: 207-10.
        [http://dx.doi.org/10.14719/pst.2016.3.2.206]

[8]   Shanker A, Bhargava A, Bajpai R, Singh S, Srivastava S, Sharma V. Bioinformatically mined simple sequence repeats in expressed sequences of *Citrus sinensis*. Sci Hortic (Amsterdam) 2007; 113: 353-61.
        [http://dx.doi.org/10.1016/j.scienta.2007.04.011]

[9]   Kabra R, Kapil A, Attarwala K, Rai PK, Shanker A. Identification of common, unique and polymorphic microsatellites among 73 cyanobacterial genomes. World J Microbiol Biotechnol 2016; 32(4): 71.
        [http://dx.doi.org/10.1007/s11274-016-2061-0] [PMID: 27030027]

[10]   Kapil A, Rai PK, Shanker A. ChloroSSRdb: a repository of perfect and imperfect chloroplastic simple sequence repeats (cpSSRs) of green plants. Database (Oxford) 2014; bau107.
        [http://dx.doi.org/10.1093/database/bau107] [PMID: 25380781]

[11]   Kumar M, Kapil A, Shanker A. MitoSatPlant: mitochondrial microsatellites database of viridiplantae. Mitochondrion 2014; 19(Pt B): 334-7.
        [http://dx.doi.org/10.1016/j.mito.2014.02.002] [PMID: 24561221]

[12]   Shukla N, Kuntal H, Shanker A, Sharma SN. Mining and analysis of simple sequence repeats in the chloroplast genomes of genus *Vigna*. Biotechnology Research and Innovation 2018; 2: 9-18.
        [http://dx.doi.org/10.1016/j.biori.2018.08.001]

[13]   Zhao CX, Zhu RL, Liu Y. Simple sequence repeats in bryophyte mitochondrial genomes. Mitochondrial DNA A DNA Mapp Seq Anal 2016; 27(1): 191-7.
        [http://dx.doi.org/10.3109/19401736.2014.880889] [PMID: 24491104]

[14]   Shanker A. Identification of microsatellites in chloroplast genome of *Anthoceros formosae*. Arch Bryol 2013; 191: 1-6.

[15]   Shanker A. Simple sequence repeats mining using computational approach in chloroplast genome of *Marchantia polymorpha*. Arctoa 2014; 23: 145-9.
[http://dx.doi.org/10.15298/arctoa.23.12]

[16]   Srivastava D, Shanker A. Identification of simple sequence repeats in chloroplast genomes of Magnoliids through bioinformatics approach. Interdiscip Sci 2016; 8(4): 327-36.
[http://dx.doi.org/10.1007/s12539-015-0129-4] [PMID: 26471998]

[17]   Kumar S, Shanker A. Common, unique and polymorphic simple sequence repeats in chloroplast genomes of genus *arabidopsis*. vegetos-. Int J Plant Res 2018; 31(special): 125-31.

[18]   Srivastava D, Shanker A. Mining of simple sequence repeats in chloroplast genome sequence of *Cocos nucifera*. Applied Research Journal 2015; 1: 51-4.

[19]   Shanker A. Sequenced mitochondrial genomes of bryophytes. Arch Bryol 2012; 146: 1-6.

[20]   Myszczyński K, Bączkiewicz A, Szczecińska M, Buczkowska K, Kulik T, Sawicki J. The complete mitochondrial genome of the cryptic species C of *Aneura pinguis*. DNA Mapp Seq Anal 2017; 28(1): 112-3.
[http://dx.doi.org/10.3109/19401736.2015.1111347] [PMID: 26678523]

[21]   Cao Y, Wang L, Xu K, *et al*. Information theory-based algorithm for *in silico* prediction of PCR products with whole genomic sequences as templates. BMC Bioinformatics 2005; 6: 190.
[http://dx.doi.org/10.1186/1471-2105-6-190] [PMID: 16042814]

[22]   Kuntal H, Sharma V. *In silico* analysis of SSRs in mitochondrial genomes of plants. OMICS 2011; 15(11): 783-9.
[http://dx.doi.org/10.1089/omi.2011.0074] [PMID: 22011339]

[23]   Shanker A. Mining of simple sequence repeats in chloroplast genome of a parasitic liverwort: *Aneura mirabilis*. Arch Bryol 2013; 196: 1-4.

[24]   Shanker A. Computational mining of microsatellites in the chloroplast genome of *Ptilidium pulcherrimum*, a liverwort. International Journal of Environment 2014; 3: 50-8.
[http://dx.doi.org/10.3126/ije.v3i3.11063]

[25]   Shanker A. Computationally mined microsatellites in chloroplast genome of *Pellia endiviifolia*. Arch Bryol 2014; 199: 1-5.

CHAPTER 8

Bryophytes as Indicators of Human Disturbances in Tropical Rain Forests

A.E. Dulip Daniels[*]

Bryology Laboratory, Department of Botany & Research Centre, Scott Christian College, (Autonomous), Nagercoil - 629 003, Tamil Nadu, India

Abstract: Bryophytes, the most primitive group of land plants, though elusive, are useful to humans in a number of ways. They are elusive due to their preference to grow in micro-habitats with high humidity. Regarding their uses, tribal people in North America and China, use bryophytes in their traditional medicines. Many have antimicrobial and anticancer properties as well. Some mosses are used as effective filtering and absorption agents in the treatment of waste water and effluents containing heavy metals and organic substances. They are widely used as indicators of heavy metals in air pollution and are also remarkable indicators of radioactive pollution. Bryophytes assist in geobotanical prospecting as ecological indicators in botanical surveys. They are generally poikilohydric, losing water rapidly with rising temperature and hence when forests are cleared or disturbed due to selective felling, these moisture and shade-loving plants perish. Therefore, bryophytes serve as good indicators of human disturbances in forests.

Keywords: Bryophytes, Indicators, Poikilohydric, Rainforests, Shade-loving.

INTRODUCTION

Bryophytes evolved about 400 million years ago [1] and are the most ancient group of land plants which probably evolved from green algal ancestors, closely related to the Charophytes [2]. The origin of bryophytes remained a topic of dispute until recent studies on cell ultra-structure and molecular biology con-firmed that this primitive group comprises three separate evolutionary lineages, the Marchantiophyta (liverworts), Bryophyta (mosses) and Anthoce-rotophyta (hornworts) [3]. However, the name, 'Bryophyte' is still used as a collective term to address these three unique groups owing to its familiarity. The term bryophyte comes from two Greek words: *'bryon'* meaning a moss and *'phyton'* a plant. It was A. Braun, who first used this term in 1864 to include the algae, fungi, lichens

[*] **Corresponding author A.E. Dulip Daniels**: Bryology Laboratory, Department of Botany & Research Centre, Scott Christian College (Autonomous), Nagercoil - 629 003, Tamil Nadu, India; Tel: +91 - 9791245551; E-mail: dulipdaniels@yahoo.co.uk

Afroz Alam (Ed.)

and mosses [4]. However, today this term is used to denote only the bryophytes. Usually bryophytes do not form a conspicuous part of any vegetation, but can be very conspicuous when they grow as extensive mats in woodlands, cushions on walls and rocks or epiphytes on the trunks and leaves of tropical rain forest trees. They are often seen as pioneer colonizers of disturbed habitats and on barren rocks after lichens. However, water is indispensable for growth and completion of their life-cycle and hence bryophytes are considered as 'amphibians' of the Plant kingdom.

Morphology and Habitat

Bryophytes differ from vascular plants in two important features. Unexceptionally, in all bryophytes the dominant phase of the life-cycle is the generally seen photosynthetic and free-living haploid gametophyte on which the short-lived diploid spore-producing sporophyte is dependent for survival though partially autotrophic when immature. The gametophyte is usually attached to the substratum by either unicellular or multi-cellular root-like hairy, branched or simple rhizoids. The sporophytes of mosses usually consist of an attachment part at the base called the foot, a simple stalk (the seta) and a single, terminal sporangium (the capsule), which produces the spores. However, this distinction is not well-defined in liverwort sporophytes. Since the sporophytes of bryophytes do not contain the special lignified water conducting xylem tissue present in the more complex sporophytes of vascular plants, bryophytes are known as non-vascular cryptogams.

Hornworts possess a thalloid gametophyte with embedded sex organs. The sporophyte is horn-like and the capsule matures in a basipetal succession. It continues to produce spores until the growing period terminates. Though the dominant phase of the life-cycle in all bryophytes is the gametophytic one, taxa are distinguished from each other based on the variations exhibited by both the gametophyte and sporophyte.

Bryophytes occupy diverse micro-habitats. Some occur exclusively in the moist, shady under-storey of least disturbed forests, and in the inner tree canopy high above the ground. Some prefer river banks and moist road-side cuts to colonize. However, generally they occur on soil and mud walls (terricolous), rocks (rupicolous), pebbles and stones (saxicolous) and decaying logs (lignicolous). Besides, growing as epiphytes on tree trunks (corticolous) and leaves (epiphyllous or folicolous) are of common occurrence. A notable feature of some bryophytes is their remarkable capacity to endure drought and are usually known as the xeromorphic forms. They form the secondary colonizers after lichens on barren rocks.

Importance of Bryophytes

Bryophytes form a significant part of the biodiversity in moist tropical forests, certain wetlands and in the tundra regions as well. Their ability to absorb and retain water is remarkable and largely contributes to the sustenance of humidity in tropical forests. The bryophyte mass in tropical montane forests is a major component of the total biomass and are therefore an important component of the hydrological, chemical and organic matter cycles [5, 6]. They are extremely good soil binders and therefore help in controlling soil erosion.

Bryophytes are quite often used as potential medicines by natives. North American Indians and Chinese use a number of bryophytes as herbal medicines to treat cardio-vascular disorders, boils, eczema, cuts, bites, wounds and burns [7 - 10]. Chemical analyses have brought to light that most bryophytes, including *Sphagnum*, have antibiotic properties [11]. Extracts of many species of mosses and liverworts contain phenolic compounds that inhibit the growth of pathogenic fungi and bacteria. Anti-cancer properties have also been reported from some liverworts [12].

Bryophtyes have important industrial uses as well. Peat has been exploited commercially for over 150 years as a source of fuel and as a soil additive. Species of *Sphagnum* are used as effective filtering and absorption agents in the treatment of waste water and effluents containing heavy metals, organic substances such as oils, detergents and dyes [13]. Bryophytes are widely used as indicators of heavy metals in air pollution, especially in large, densely populated cities and in areas around power stations and metallurgical works [14, 15]. They are also remarkable indicators of radio-active pollution and are used in monitoring radioactive caesium [16]. Bryophytes assist in geobotanical prospecting [17] as ecological indicators in botanical surveys. For instance, the mosses *Tortella tortuosa* (Schrad. ex Hedw.) Limpr., and *Racomitrium lanuginosum* (Hedw.) Brid. grow on calcareous and acidic substrates, respectively.

Bryophyte communities are a haven for microorganisms, rotifers, nematodes, earthworms, molluscs, insects, spiders and many other invertebrates which include millipedes and earthworms [18, 19]. Bryophytes provide food, shelter and nesting material for small animals and invertebrates and indirectly serve as a matrix for interaction among the organisms. Some species are used as nesting materials by birds [19 - 21]. For example, in tropical montane forests, pendant or trailing mosses, especially *Cryptopapillaria fuscescens* (Hook.) M. Menzel, *Floribundaria floribunda* (Dozy & Molk.) M. Fleisch., *Meteorium polytrichum* Dozy & Molk. and a number of liverworts such as *Frullania ericoides* (Nees) Mont. and *Plagiochila sciophila* Nees ex Lindenb. are used to camouflage their nests. Bryophytes also provide favourable substrata as seed beds for the establishment of seedlings of higher plants [19].

Bryophytes in Tropical Forests

Two thirds of all bryophyte species occur in the tropical region, especially in evergreen forests [22 - 24]. Tropical rain forests are unique ecosystems existing on earth for hundreds of millions of years. Most tropical rain forests are fragments of Gondwanaland, the supercontinent of the Mesozoic era [25]. Despite the fact that, Tropical rain forests occupy only 7% of the land surface, they hold half of the species of the Earth. The abundance of epiphytic bryophytes is a special character of tropical rain forests [26]. On the contrary, terricolous forms are rare, which is probably due to the large amount of litter and its fast decomposition. However, roadside banks and cuts are welcome habitats for terricolous bryophytes with increasing elevation. The lower temperatures and comparatively higher light levels in cloud forests and the availability of plentiful water due to mist and fog favour the abundant growth of epiphytic and terrestrial forms in tropical rain forests [23, 27].

In India, tropical forests are found in the Andaman and Nicobar Islands, the Western and Eastern Ghats in Peninsular India and the North-east.

Like any other tropical forest, the forests of the Western and Eastern Ghats suffered severe human disturbances such as clearing of vast tracts of rich evergreen and moist deciduous forests for agriculture, introduction of monoculture plantations such as pine, eucalyptus, cinchona, wattle, tea, coffee *etc.* by the Colonial Europeans and later as rubber, cocoa, betel leaves, tobacco, clove, pepper, areca nut, nutmeg, cardamom, *etc.* by the private sectors and the Forest Departments as well for more than two centuries [28]. Besides, the construction of dams and reservoirs, roads, factories, hospitals, educational institutions, residential buildings *etc.*, added to their woes. These interferences have significant effects on the biodiversity as a whole, especially the delicate and primitive moisture-loving plants, the bryophytes [29, 30].

Bryophytes are generally poikilohydric in nature and hence lose water rapidly with rising temperature. Therefore, when forests are cleared or disturbed due to selective felling, these moisture and shade-loving plants get exposed to direct sunlight, which directly affects their growth and life-cycle, and when exposed to light and desiccation for a long period of time, they ultimately perish. Therefore, how can bryophytes serve as indicators of human disturbances in forests?

Indicator Species

Studies and observations made on the diversity and occurrence of the bryophytes of the Western and Eastern Ghats for the past twenty years show that, mosses such as *Actinodontium ascendens* Schwägr., *A. rhaphidostegum* (Müll.Hal.) Bosch & Sande Lac., *Callicostella papillata* (Mont.) Mitt., *Cyathophorum adiantum* (Griff.) Mitt., *C. hookerianum* (Griff.) Mitt., *Daltonia angustifolia* Dozy & Molk., *D. contorta* Müll. Hal., *Diphyscium fasciculatum* Mitt., *D. mucronifolium* Mitt. *Distichophyllum ceylanicum* (Mitt.) Paris, *D. decolyi* Gangulee, *Fissidens angustifolius* Sull., *F. anomalus* Mont., *F. firmus* Mitt., *Funaria hygrometrica* Hedw., *Groutiella tomentosa* (Hornsch.) Wijk & Margad., *Homaliadelphus targionianus* (Mitt.) Dixon & P. de la Varde, *Hookeriopsis secunda* (Griff.) Broth., *Lepidopilidium furcatum* (Thwaites & Mitt.) Broth., *Neckera himalayana* Mitt., *Physcomitrium eurystomum* Sendtn., *Thamniopsis utacamundiana* (Mont.) W.R. Buck, *Macromitrium* spp., *Pterobryopsis* spp., *Symphyodon* spp., *Symphysodontella* spp., *Hypopterygium* spp., *Leucobryum* spp. and most of the Hypnobryales grow exclusively in montane evergreen forests and/or wet evergreen forests which have closed canopies with abundant moisture content. Any disturbance that permits penetration of light and a rise in temperature would grossly affect the survival of these delicate species. Hence, the presence of any of these species in an evergreen forest would indicate that the particular patch is either a least disturbed one or highly degraded with only relict species of these delicate mosses. However, presence of higher number of these species would indicate that the forest is a least disturbed one and lower number would indicate that the forest is degraded but still has enough potential to sustain such delicate mosses [31].

Mosses such as *Meteoriopsis ancistrodes* (Renauld & Cardot) Broth., *Meteoriopsis reclinata* (Müll. Hal.) M. Fleisch., *Toloxis semitorta* (Müll. Hal.) W.R. Buck, *Calyptothecium recurvulum* (Broth.) Broth., *C. wightii* (Mitt.) M. Fleisch., *Neckeropsis gracilenta* (Bosch & Sande Lac.) M. Fleisch., *Racopilum cuspidigerum* (Schwägr.) Ångstr., *R. orthocarpum* Wilson *ex* Mitt., *etc.* and most *Fissidens* spp. have the adaptability to grow in moist deciduous forests, degraded forests and even in plantations, and therefore known as 'opportunists'. The presence of any of these species would indicate that the patch of forest is a degraded one paving way for the invasion of these opportunists which are robust and aggressive colonizers and ultimately replace the more delicate and moisture-loving species.

Members of Pottiaceae are highly adaptable and can grow in any harsh habitat. Most of the light tolerant and drought enduring species such as *Barbula indica* (Hook.) Spreng., *B. javanica* Dozy & Molk., *Gymnostomiella vernicosa* (Hook.

ex Harv.) M. Fleisch., *Hymenostylium recurvirostrum* (Hedw.) Dixon var. *recurvirostrum*, *H. recurvirostrum* (Hedw.) Dixon var. *aurantiacum* (Mitt.) Gangulee, *Hyophila involuta* (Hook.) A. Jaeger and *Weissia edentula* Mitt. belong to this family. Hence, the presence of any of these species would indicate that the patch of forest under study is a highly disturbed one paving way for the colonization of light tolerant and drought enduring species, and not suitable for the growth of delicate and exclusive species.

Similarly, liverworts such as *Anastrophyllum aristatum* (Herzog ex N. Kitag.) A.E.D. Daniels & al., *Chandonanthus hirtellus* (F. Weber) Mitt., *Cephaloziella capillaris* (Steph.) Douin., *Frullania apiculata* (Reinw. & al.) Dumort., *F. arecae* (Spreng.) Gottsche, *F. berthoumieui* Steph., *F. densiloba* Steph. ex A. Evans, *F. neurota* Taylor, *F. ramuligera* (Nees) Mont., *F. riojaneirensis* (Raddi) Ångstr., *Heteroscyphus perfoliatus* (Mont.) Schiffn., *Lophocolea muricata* (Lehm.) Nees., *Herbertus armitanus* (Steph.) H.A. Mill., *H. dicranus* (Taylor ex Gottsche & al.) Trevis., *Jungermannia infusca* (Mitt.) Steph., *Cheilolejeunea trifaria* (Reinw. & al.) Mizut., most species of *Cololejeunea, Lejeunea flava* (Sw.) Nees, *Lopholejeunea subfusca* (Nees) Schiffn., *Ptychanthus striatus* (Lehm. & Lindenb.) Nees, *Thysananthus spathulistipus* (Reinw. & al.) Lindenb. *Mastigophora diclados* (Brid. ex F. Weber) Nees ex Schiffn. *Plagiochilion oppositum* (Reinw., Blume & Nees) S. Hatt. *Schistochila aligera* (Nees & Blume) J.B. Jack & Steph. and *Dumortiera hirsuta* (Sw.) Nees grow in montane evergreen forests and/or wet evergreen forests. The presence of any of these species in an evergreen forest would indicate the health of the forest and that the particular patch is a least disturbed one, whereas the presence of *Frullania acutiloba* Mitt., *F. inflexa* Mitt., *F. muscicola* Steph., *F. campanulata* Sande Lac., *F. ericoides* (Nees) Mont., *Jungermannia tetragona* Lindenb., *Plagiochasma rupestre* (J.R. Forst. & G. Forst.) Steph., *Exormotheca ceylonensis* Meijer, *Riccia billardierei* Mont. & Nees, *Targionia hypophylla* L. and *Cyathodium* spp. are indicative of unhealthy and degraded forests with increasing dryness and light intensity that would ultimately annihilate the delicate and exclusive species.

Hornworts, though delicate and moisture-loving, are quite adapted to growing in microclimatic niches such as small shady water sources, under or adjacent to fallen logs in open degraded forests and plantations as well. Hence, they do not serve as good indicators of forest health.

Besides the above mentioned indicator species, there are several other species of mosses and liverworts that serve the purpose. However, the more common ones have been discussed here.

CONSENT FOR PUBLICATION

Not applicable.

CONFLICT OF INTEREST

The authors confirm that this chapter contents have no conflict of interest.

ACKNOWLEDGEMENTS

Declared none.

REFERENCES

[1]   Qiu YL, Palmer JD. Phylogeny of early land plants: Insights from genes and genomes. Trends Plant Sci 1999; 4(1): 26-30.
[http://dx.doi.org/10.1016/S1360-1385(98)01361-2] [PMID: 10234267]

[2]   Shaw J, Renzaglia K. Phylogeny and diversification of bryophytes. Am J Bot 2004; 91(10): 1557-81.
[http://dx.doi.org/10.3732/ajb.91.10.1557] [PMID: 21652309]

[3]   Konrat MV, Shaw AJ, Renzaglia KS. The closest living relatives of early land plants. Phytotaxa 2010; 9: 5-10.
[http://dx.doi.org/10.11646/phytotaxa.9.1.3]

[4]   Braun R. Uebersicht des naturlichen.Flora der Provinz Brandenburg. Berlin 1864.

[5]   Hofstede RGM, Wolf JHD, Benzing DH. Epiphytic biomass and nutrient status of a Colombian upper montane rain forest. Selbayana 1993; 14: 37-45.

[6]   Rhoades F. Nonvascular epiphytes in forest canopies: Worldwide distribution, abundance and ecological roles. Forest Canopies. San Diego: Academic Press 1995; pp. 353-408.

[7]   Flowers S. Ethnobryology of the Gosuite Indians of Utah. Bryologist 1957; 60: 11-4.
[http://dx.doi.org/10.1639/0007-2745(1957)60[11:EOTGIO]2.0.CO;2]

[8]   Wu PC. *Rhodobryum giganteum* (Schwaegr.) Par. can be used for curing cardiovascular disease. Acta Phytotax Sin 1977; 15: 93.

[9]   Ando H. Use of bryophytes in China. Proc Bryol Soc Japan, Tokyo 1983; 3: 104-6.

[10]   Ding H. Medicinal spore-bearing plants of China. Shanghai, China: Shanghai Science and Technology Press 1982.

[11]   Banerjee RD. Studies on antibiotic activity of bryophytes and pteridophytes 1974. unpublished.

[12]   Spjut RW, Cassady JM, McCloud T, *et al.* Variation in cytotoxicity and antitumour activity among samples of the moss *Claopodium crispifolium* (Thuidiaceae). Econ Bot 1988; 42: 62-72.
[http://dx.doi.org/10.1007/BF02859034]

[13]   Poots VJP, McKay G, Healy JJ. The removal of acid dye from effluent using natural adsorbents I. peat. Water Res 1976; 10: 106-1066.
[http://dx.doi.org/10.1016/0043-1354(76)90036-1]

[14]   Maschke J. Moose als Bioindikatoren von Schwermetall Immissionen. Bryophyt Bibl 1981; 22: 492.

[15]   Mäkinen A. Use of Hylocomium splendens for regional and local heavy metal monitoring around a coal-fired power plant in southern Finland. Symp Biol Hung 1987; 35: 777-94.

[16]   Isomura K, Higuchi M, Shibata H, Tsukada H, Iwashima K, Sugiyama H. Distribution of radio-active caesium in mosses and application of mosses for monitoring of radio-active caesiums. Radioisotopes

1993; 42: 157-63.
[http://dx.doi.org/10.3769/radioisotopes.42.157]

[17]   Shacklette HT. The use of aquatic bryophytes in prospecting. J Geochem Explor 1984; 21: 89-93.
[http://dx.doi.org/10.1016/0375-6742(84)90036-0]

[18]   Gerson T. Bryophytes and invertebrates. Bryophyte Ecology. London, New York: Chapman & Hall 1982; pp. 291-332.
[http://dx.doi.org/10.1007/978-94-009-5891-3_9]

[19]   Daniels AED, Daniel P. The Bryoflora of the Southernmost Western Ghats, India. Dehra Dun, India: Bishen Singh Mahendra Pal Singh 2013.

[20]   Pant GB, Tewari SD. Birds gather bryophytes for nest building. Phyta J Soc P. Taxonomists 1981-1982; 1984(5): 57-60.

[21]   Longton RE. The role of bryophytes and lichens in terrestrial ecosystems.Bryophytes and lichens in a changing environment. Oxford, England: Clarendon Press 1992; pp. 234-58.

[22]   Pócs T. Tropical forest bryophytes.Bryophyte Ecology. London: Chapman & Hall, England 1982; pp. 59-105.
[http://dx.doi.org/10.1007/978-94-009-5891-3_3]

[23]   Richards PW. The ecology of tropical forest bryophytesNew Manual of Bryology. Japan: Hattori Botanical Laboratory: edited by Schuster RM, Published. Nichinan 1984; II: pp. 1233-70.

[24]   Gradstein SR. The vanishing tropical rain forest as an environment for bryophytes and lichens.Bryophytes and lichens in a changing environment. Oxford, England: Clarendon Press 1992; pp. 236-58.

[25]   Corlette R, Primack R. Tropical rain forests and the need for cross-continental comparisons.Trends Ecol Evol 2006; 21-104.

[26]   Ramsay HP, Cairns A. Habitat, distribution and the phytogeographical affinities of mosses in the Wet Tropicos bioregion, north-east Queensland, Australia. Cunninghamia 2004; 8: 371-408.

[27]   Gradstein SR, Pócs T. Biogeography of tropical rain forest bryophytes. Tropical Rain Forest ecosystems (Series Ecosystems of the World Vol 14A). Amsterdam, The Netherlands: Elsevier Science Publisher 1989; pp. 311-25.
[http://dx.doi.org/10.1016/B978-0-444-42755-7.50022-5]

[28]   Daniels AED, Kariyappa KC, Mabel JL, Raja RDA, Daniel P. The overlooked and rare liverwort *Frullania ramuligera* (Nees) Mont. (*Frullaniaceae*) rediscovered in the Western Ghats of India. Acta Bot Hung 2010; 52: 297-303.
[http://dx.doi.org/10.1556/ABot.52.2010.3-4.8]

[29]   Daniels AED. Introduction of monoculture plantations and their impact on the wildlife of Kanyakumari Dist. Zoos' Print 1998; 13: 16-8.

[30]   Kariyappa KC, Daniels AED. Seasonal variations influencing the bryophyte diversity of monoculture plantations in the Southern Western Ghats. In: Laladhas KP, Oommen OV, Sudhakaran PR, Eds. Biodiversity Conservation - challenges for future. UAE: Bentham Science Publishers. 2015; pp. 223-8.

[31]   Alam A. Bryomonitoring of Environmental Pollution In: Biotic and Abiotic Stress Tolerance in Plants. Sigapore, Springer Nature: Edited By Vats S. 2018; pp. 349-65.
[http://dx.doi.org/10.1007/978-981-10-9029-5_13]

CHAPTER 9

An Account of Genus *Porella* (Dill.) L. (Porellaceae, Marchantiophyta) in Nilgiri Hills, Western Ghats

Praveen Kumar Verma[*]

Forest Botany Division, Forest Research Institute, Dehradun, Uttarakhand, India

Abstract: The taxa of genus *Porella* (Dill.) L. in Nilgiri Hills, Western Ghats are discussed, keyed, illustrated and provided with distribution. These are *Porella madagascariensis* (Nees & Mont.) Trevis, *P. chinensis* Steph., *P. perrottetiana* (Mont.) Trevis., *P. acutifolia* (Lehm. & Lindenb.) Trevis., *Porella caespitans* S. Hatt. var. *setigera* (Steph.) S. Hatt., *P. campylophylla* (Lehm. & Lindenb) Trevis. and *P. campylophylla* (Lehm. & Lindenb.) Trevis. var. *ligulifera* (Tayl.) S. Hatt., and one more species of *Porella*, *P. pinnata* was reported earlier from the area but could not be collected from any of the explored localities of the Nilgiri hills of Tamil Nadu. Illustrations of all the species of *Porella* recorded are also provided.

Keywords: Western Ghats, Tamil Nadu, Marchantiophyta, Porellaceae, *Porella*.

INTRODUCTION

The family Porellaceae is characterized by the robust size of plants, non rostellate perianth and lack of strong specialized leaf-lobule. Only 3 representatives of this family viz., *Ascidiota* Mass. (Asian–North American), *Macvicaria* Nicholas (Asian) and *Porella* (Dill.) Linn. (cosmopolitan) is reported so far [1]. Members of the family are widely distributed on the globe, especially in both tropical and temperate regions. Out of the 3 genera only *Porella* is known from India as well as in the Nilgiri hills.

Recently, Verma *et al.* [2, 3] reported 326 taxa of bryophytes from Nilgiri hills, in which 157 taxa under 81 genera of Musci and 169 taxa belonging to 58 genera of Hepaticae and Anthocerotae including 7 taxa of genus *Porella*. The generic delimitation of the genus is mainly based on foliar (leaf-lobe, leaf-lobule and underleaf) morphology, and branching pattern. Most of the species of *Porella* found in the Nilgiri hills are pinnately branched, the branching pattern *Frullania-*

---

[*] **Corresponding author Praveen Kumar Verma:** Forest Botany Division, Forest Research Institute, Dehradun, Uttarakhand, India; Tel: +91-7579422246; E-mail: pkverma_bryo@yahoo.co.in

Afroz Alam (Ed.)

Ptychanthus type (in which the first under-leaf of branch is bilobed). Stem anatomy is stable and more or less similar in all species and doesn't seen any taxonomic value, but leaves and under-leaves shows phenotypic plasticity and are of great taxonomic value, ranging from closely imbricate to distant (contiguous), entire to dentate, and differently shaped. The taxonomic parameters used for species determination are plant size; stem anatomy; leaf orientation and arrangement, shape, size, margin, number of dentition and type of cells of leaf-lobe, leaf–lobule and under-leaves, leaf decurrence and sexuality (if present).

The genus *Porella* is one of the largest among the suborder Porellineae. Steere and Schuster [4] reported 153 binomials across the globe while Qian *et al.* [5] reported 86 species, mostly distributed in tropical to subtropical parts of the world. In Asia this genus grows with luxuriance and more than 200 binomials were earlier reported from this continent. However, Hattori [6] reduced the number and reported 58 genuine species from the region. Most of the taxa were either synonymies under different names or even subjected to changes in generic ranks. In India the genus is represented by 27 taxa [7]. In Nilgiri hills a number of binomials were reported from time to time by different workers, which are now reduced to 8 valid taxa (7 species and one variety). It is one of the most frequent occurring genera in Nilgiri hills and occupying almost every habitat *viz.*, corticolous, lignicolous and terricolous.

The preliminary report of the genus from the Nilgiri hills comes through the publication of Montagne [8], who reported *Madotheca perrottetiana* Mont. [now *Porella perrottetiana* (Mont.) Trevis.] and *M. nilgheriensis* Nees & *t* Mont. [now *P. madagascariensis* (Nees & Mont) Trev.]. Subsequently Mitten [9] added *Madotheca acutifolia* Lehm. & Lindenb. [now *P. acutifolia* (Lehm. & Lindenb.) Trevis.] and *M. ligulifera* Taylor [now *P. campylophylla* (Lehm. & Lindenb.) Trevis. var. *ligulifera* (Tayl.) S. Hatt.] from the region. After a long gap Stephani [10] reported *Madotheca calcarata* Steph., [now *Porella caespitans* S. Hatt. var. *setigera* (Steph.) S. Hatt.] from Nilgiri hills. Further, Chopra [11] published two species of *Porella*, as *Madotheca campylophylla* Lehm. & Lindenb [now *P. campylophylla* (Lehm. & Lindenb) Trevis.] and *M. porella* (Dick.) Nees [now *P. pinnata* (Dick.) Lindenb.] from Kotagiri. Out of these *P. pinnata* could not be collected from any of the explored localities of the Nilgiri hills.

In the present work, *Porella chinensis* Steph. is reported here as new record from the south Indian territory (Ootacamund – Love Dale), resulting into an increase in the bio-diversity status of the genus in the area.

MATERIALS AND METHODS

The Nilgiri or 'Blue Mountains' owe its name due to the predominant and verdant blue blooms of an angiosperm – *Strobilanthes kunthiana* (family Acanthaceae). It is situated at 10°1'-11°45' N latitude, 76°-77°15' E longitude, and about 1200-2500 m altitude spread over an expanse of 2600 km² forming the centre of diversification of bryophytes in the India's oldest Biosphere Reserve, the Nilgiri Biosphere Reserve (NBR), of Tamil Nadu.

Fresh as well as herbarium specimens collected from different part of Nilgiri hills (Tamil Nadu) by the author himself and collections made by the late Prof. S.K. Pande, late Prof. R. Udar, Prof. S. C. Srivastava and their associates available in Bryophyte Herbarium, Department of Botany, University of Lucknow, Lucknow (LWU) were examined during the present investigation. Several type and authenticated specimens related to the liverworts of Nilgiri hills located in different reputed herbaria and exsiccatae of the world, including Conservatoire *et* Jardin Botaniques, Geneva (G) through the courtesy of the curator who very kindly loaned these specimens to me through Prof. S.C. Srivastava (the guide of the author) for examination. All line drawings were drawn by the author himself with the help of Camera Lucida (Nikken – Japan).

OBSERVATIONS

Key to the Species of Genus *Porella* (Dill.) L. in Nnilgiri hills

1. Plants small to medium sized (30-60 mm); leaf-lobes rounded, sub-truncate, obtuse at apex. …..………….…..………………………………..…….……….. 2
    1. Plants large (60-130) mm; leaf-lobe acute or sub-acute to dentate at apex………….………..…………….…..……………………………………… 3
    2. Plants smooth, sparsely branched; leaves distant; margins of leaf-lobes and leaf-lobule crispate ……………..………………………………...…*P. madagascariensis*
2. Plants flaccid; pinnately branched; leaves imbricate; margins of leaf-lobes and leaf lobule smooth or entire, sometimes slightly crispate…………*P. chinensis*
3. Leaf-lobe, leaf-lobules and underleaves spinosely toothed throughout the margin, tooth up to 15 (18) cells long. ……… *P. perrottetiana*
    1. Leaf-lobe, leaf-lobules, underleaves acute to sub-acute (acuminate), irregularly toothed, tooth up to 1-5 (7) cells long ……... …………..…4
    2. Leaf-lobe ovate-triangular, with pilose apex or rarely truncate to obtuse and minutely toothed at apex; apical teeth 1-2 …………………………… 5
4. Leaf-lobe wide, ovate to oblong-ovate, apex sub-truncate, obtuse, rounded, numerously toothed, apical teeth 3-11 …………………………… 6
5. Leaf-lobe ovate, oblong, with dentate apex with 1-2 celled minute teeth, rarely with one additional tooth, leaf-lobule obtuse to sub-acute or acuminate. …………………... *P. acutifolia*

1. Leaf-lobe ovate, triangular with pilose apical tooth, tooth 5-9 cells large, triangulate;   leaf-lobule   obtuse-sub   truncate   or setose..*P. caespitans* var. *setigera*

2. Leaf-lobe apex with few, short, sub-apical tooth, underleaves with obtuse apex, or rarely dentate apex.. *P. campylophylla*

6. Leaf-lobe apex with numerous acute apical teeth, underleaves with truncate, with dentate apex ....…...........*P. campylophylla* var. *ligulifolia*

Porella madagascariensis (Nees & Mont.) Trevis.; *Mem. Real Istit. Lombardo Sci Lett.* ser. 3, **4**: 407 (1877). – *Madotheca nilgheriensis* Mont., *Ann. Sci. Nat.* ser. 2, **17**: 15 (1842). – *Porella nilgheriensis* (Mont.) Trevis., *Mem. Real Istit. Lombordo Sci. Lett.* ser. 3, **4**: 408 (1877) (Fig. **1**: 1-17).

Plants prostrate, light green in colour, up to 45 mm long, 1.8-2.0 mm wide, sparsely branched, branching '*Frullania*-type'. Stem 15-17 cells across the diameter, differentiated, cortical cells in 2 rows, small, thick-walled, 9-15×9-15 μm, isodiametric, medullary cells large, thin-walled with indistinct trigones, 21-36×21-30 μm. Leaves contiguous (or slightly imbricate), horizontally spreading, at an angle of 80-90°; leaf -lobe oblong, 0.63-1.2 mm long, 0.8-1.0 mm wide, apex rounded-oblong (truncate), entire with crispate margin, sometime 1-2 teeth present at apex; apical cells 14-17×14-16 μm, trigones small, median cells 21-38×16-21 μm, basal cells 35-42×18-36 μm; leaf - lobule parallel to stem, lanceolate, appressed, 0.30-0.39 mm long, 0.16-0.21 mm wide, apex truncate, outer base with 1(2) teeth, inner base short decurrent, appendiculate, and dentate. Underleaves contiguous, ovate-rectangulate, 0.28-0.36 mm long, 0.12-0.21 mm wide, entire, apex truncate, base long decurrent. Plants sterile.

Type Locality: Madagascar (the Republic of Madagascar) [12].

Range: Ethiopian region: Africa- Madagascar. Oriental region: Asia - Afghanistan, INDIA, Indo-China (Laos, Vietnam, Cambodia), China, Sri Lanka [6, 12, 19].

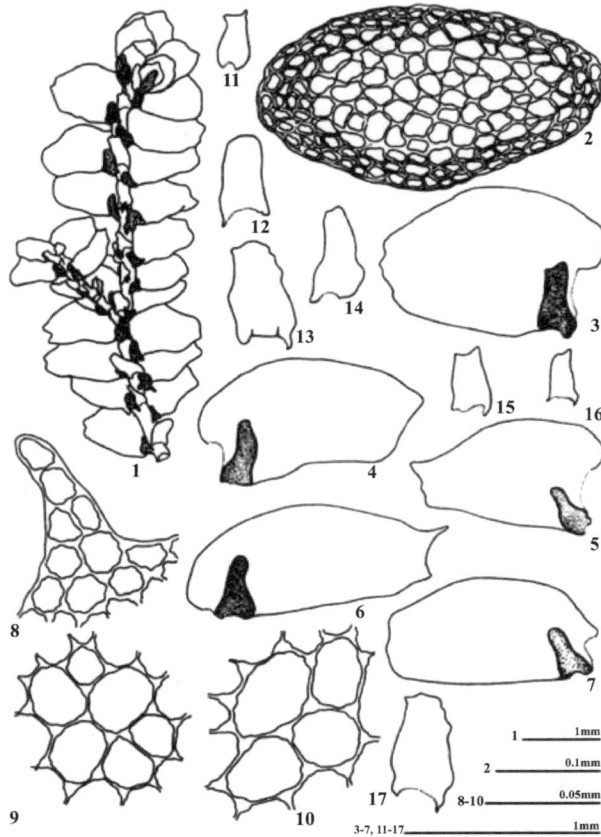

Fig. (1). 1-17: *Porella madagascariensis* (Nees & Mont) Trev. 1. Plant, ventral view, 2. Cross section of stem, a portion, 3-7. Leaves, 8. Apical cells of leaf-lobe, 9. Median cells of leaf-lobe, 10. Basal cells of leaf –lobe, 11-17. Underleaves (All figures drawn from LWU 12562/2000).

Distribution in India: Western Himalayas: Himachal Pradesh – Kullu. South India: Tamil Nadu – Nilgiri hills [Coonoor, Upper Bhavani (Avalanche)]; Palni hills (Shembaganur) [12].

Ecology: Plants growing in smooth mats in diffuse patches as the epiphytic population on the basal part of the trees (angiosperms and gymnosperms) especially on decaying logs.

Characteristics of the Species: 1. Leaves contiguous (slightly imbricate), entire 2. Leaf-apex oblong to rounded truncate (often with 1-2 short dentitions) 3. Leaf-lobule lanceolate, apex some hat obtuse 4. Outer base of leaf-lobule with 1 (2) teeth.

Specimen Examined: South India: Tamil Nadu: Nilgiri Mt. Perrottet; as *Madotheca nilgheriensis* Mont., G 009640 (Type). Tamil Nadu: Nilgiri hills – Upper Bhavani (Avalanche); *ca* 2250 m; 02.06.1972; R. Udar and party; 71 S/1972 (LWU). Upper Bhavani (Avalanche); *ca.* 2250 m; 09.10.2000; S.C. Srivastava and party; 12557/2000, 12562/2000, 12596/2000 (LWU).

Porella madagascariensis was reported from the Nilgiri hills by Montagne in 1842 as *Madotheca nilgheriensis* but the latter was synonymzed under the former [12]. The species is very rare in distribution and was recently collected from Avalanche in Nilgiri hills.

The plants mostly have entire leaves, but some of the apical leaves and branch leaves are sometimes dentate at margin. The species is closely allied to another West Himalayan species of Porella, *P. hattorii* Udar et Shaheen in shape and size of plant, leaf-lobule and underleaf morphology, but slightly differ in leaf margin, *P. hattorii* has acute to dentate leaf apex [21].

Porella chinensis (Steph.) S. Hatt., *Journ. Hattori Bot. Lab.* 30: 31 (1967) (Fig. **2**: 1-15)

Plants prostrate, soft in texture, light green, light brown (in the herbarium), up to 65 mm long, 1.8-2.2mm wide, pinnately branched, branching '*Frullania*-type'. Stem 17-20 cells across the diameter, differentiated, cortical cells thick-walled, 11-19×7-15 µm, medullary cells large, thin-walled, (11) 20-34 × (11) 15-22 µm. Leaves sub-imbricate, horizontally spreading, obliquely inserted, dorsal leaf-lobe spreading at an angle of 65°, ovate, 1.2-1.7 mm long, 0.88-1.2 mm wide, margins recurved, smooth or entire, sometimes slightly crispate, obtuse, base long decurrent; apical cells 15-22×8-15 µm, without trigones, median cells 30-38×1--26 µm, trigones acute, basal cells 26-34×11-19 µm, polygonal; leaf-lobule ligulate, apex obtuse (sub-acute), margin slightly recurved, the ventral base longly decurrent. Under-leaves imbricate, transversely inserted, ligulate, obtuse, 0.64-0.88 mm long, 0.58- 0.68 mm wide, sub-truncate, margin slightly incurved, base shortly decurrent. Plant sterile.

Type Locality: China- Shen-Si [12].

Range: Oriental region: Asia - China, INDIA. Palearctic region: Asia- Russia (Siberia -Amur, Sakhalin) [6, 12].

Distribution in India: Western Himalayas: Jammu and Kashmir – Ravi valley; Himachal Pradesh (Chamba, Simla, Rohtang, pass, Jalori pass, Great Himalayan National Park; Uttarakhand –Kumaon, Bhagirathi, Nainital. South India: Tamil Nadu – Nilgiri hills: Ootacamund - Love Dale [13 - 15].

Ecology: Plants growing in fan like forms as the epiphytic population on base of *Eucalyptus* tree.

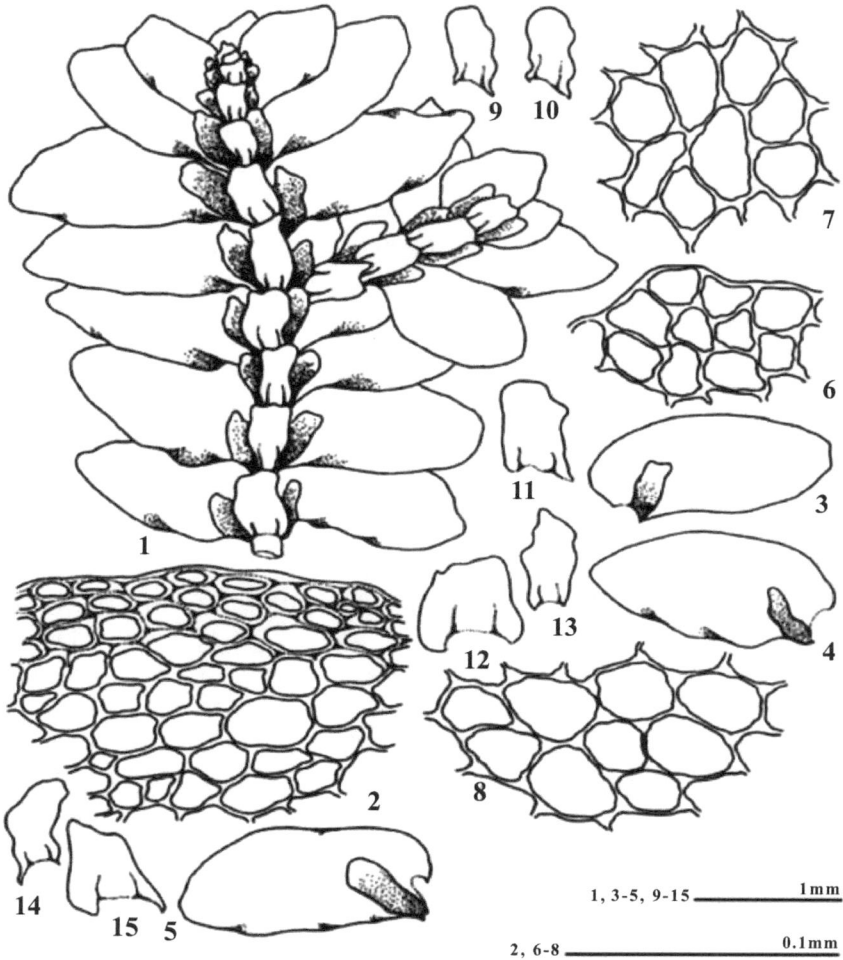

Fig. (2). 1-15: *Porella chinensis* (Steph.) S. Hatt.. 1. Plant, ventral view, 2. Cross section of stem, a portion, 3-5. Leaves, 6. Apical cells of leaf-lobe, 7. Median cells of leaf-lobe, 8. Basal cells of leaf –lobe, 9-15. Underleaves (All figures drawn from LWU 14628/2001).

Characteristics of the Species: 1. Plants flaccid 2. Leaves in loose texture, slightly crispate 3. The absence of marginal teeth in leaves and underleaves.

Specimen Examined: South India: Tamil Nadu: Nilgiri hills – Ootacamund (Love Dale); *ca.* 2250 m; 30.11.2001; P.K. Verma and A. Alam; 14628/2001 (LWU).

The species was earlier known from Eastern and Western Himalayas only [13] and is being reported as a new addition to south Indian territory though rare.

***Porella perrottetiana* (Mont.) Trevis.,** *Mem. Real Istit. Lombardo s*er. 3, 4: 408 (1877); – *Madotheca perrottetiana* Mont. *Ann. Sci. Nat. s*er. 2, 17: 15 (1842) (Fig. **3**: 1-26).

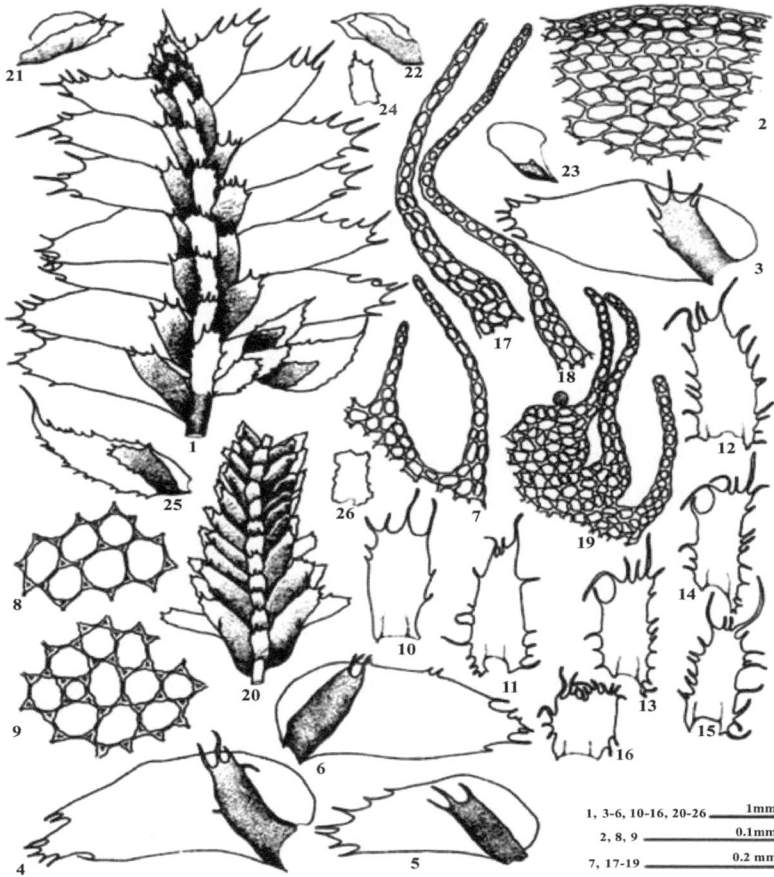

Fig. (3). 1-26: *Porella perrottetiana* (Mont.) Trevis. 1. Plant, ventral view, 2. Cross section of stem, a portion, 3-6. Leaves, 7. Apical cells of leaf-lobe, 8. Median cells of leaf-lobe, 9. Basal cells of leaf –lobe, 10-16. Underleaves, 17, 18. Apical cells of underleaf, 19. Basal cells of underleaf, 20. Androecial branch, 21-23. Male bract, 24. Male bracteole, 25. Female bract, 26. Female bracteole (All figures drawn from LWU 16960/2003).

Plants prostrate, dark green-pale green, up to 140 mm long and 7.0-8.0 mm wide, bipinnately branched, branching '*Frullania*-type'. Stem 18-26 cells across the diameter, differentiated, cortical cells small, 3 layered, walls thickened, 6-25×1--25 µm, medullary cells 13-36×13-52 µm, thin-walled. Leaves densely imbricate, obliquely inserted, horizontally spreading; leaf-lobes oblong-ovate, margins highly dentate, 3.0-5.2 mm long with dentitions, 1.3-2.7 mm wide, apex acumi-nate to acute, dentitions up to 22 cells long and 3 cells wide at base, dorsal margin arched, ventral margin straight, base decurrent; apical marginal cells 7-32×10-36 µm, trigones small, median cells 23-48×24-36 µm, polygonal, trigones well-developed, basal cells 20-55×13-30 µm, trigones well developed, base decurrent. Underleaves densely imbricate, oblong, 1.0-2.8 mm long, 0.4-1.1 mm wide, margins highly dentate, apex obtuse, base longly decurrent.

Dioicous. Androecia terminal on short lateral branches, also intercalary in position, spikate; male bracts small, delicate, densely saccate, minutely dentate. Gynoecia intercalary, on main axis, immature; female bracts in single pair, 2.6-2.8x0.76-1.0 mm, margin dentate; bract-lobule 1.0-1.6×0.44-0.52 mm; bracteole one, 1.2-0.8 mm, dentate. Sporophyte immature.

Type Locality: Tamil Nadu-Nilgiri hills [12].

Range: Oriental region: Asia - Bhutan, INDIA, Indo-China (Cambodia, Vietnam, Laos), Myanmar, Philippines, Sri Lanka, Taiwan. Palearctic region: Asia -Japan, Northern China [6, 12, 16 - 18].

Distribution in India: Eastern Himalaya: Manipur- Senapati; South India: Kerala: Anamudi Shola National Park; Tamil Nadu - Nilgiri hills [Coonoor, Ootacamund (Dodabetta, Glenmorgan, Government Botanical Garden), Kotagiri, Upper Bhavani (Avalanche)]; Palni hills- Kodaikanal (Shembaganur) [7, 11, 20, 21].

Ecology: Plants growing in fan like forms with robust patches as epiphytic population, especially on the middle tree trunk as well as primary and secondary branches of big trees, occasionally found on extremely wet rocks near the water channels.

Characteristics of the Species: 1. Plants robust, up to 140 mm long 2. Leaf-lobes oblong-ovate, margin highly dentate with acute-acuminate apex 3. Leaf-lobules, under-leaves oblong with spinosely toothed margins.

Specimens Examined: South India: Tamil Nadu: Nilgiris; ca. 1400; Perrottet; as *Madotheca perrottetiana*; 21350 (G). Ootacamund – Upper Bhavani (Avalanche); ca. 2250 m., 09-10.2000; S.C. Srivastava and party; 12545/2000, 12557/2000, 12562/2000, 12596/2000 (LWU). Ootacamund (Glenmorgan); ca. 2200 m.; 02.04.2003; P.K. Verma and A. Alam; 16960/2003, 16972/2003 (LWU).

Porella perrottetiana was reported by Montagne (1842) as *Madotheca perrottetiana* from Nilgiri hills. The species is largest among Indian species of *Porella* (up to 140 mm). The plants shrinking plasticity in morphology and number of dentitions on leaves. In the recent past collection the species was collected from Glenmorgan Avalanche and Dodabetta (altitude between 2200-2600 m) and is easily recognizable from other Indian species of the genus by its dentitions on leaf-lobe, leaf-lobule and under-leaf margins.

***Porella acutifolia* (Lehm. & Lindenb.) Trev.,** *Mem. Real Istit. Lombardo* ser. 3, 4: 408 (Fig. **4**: 1-8).

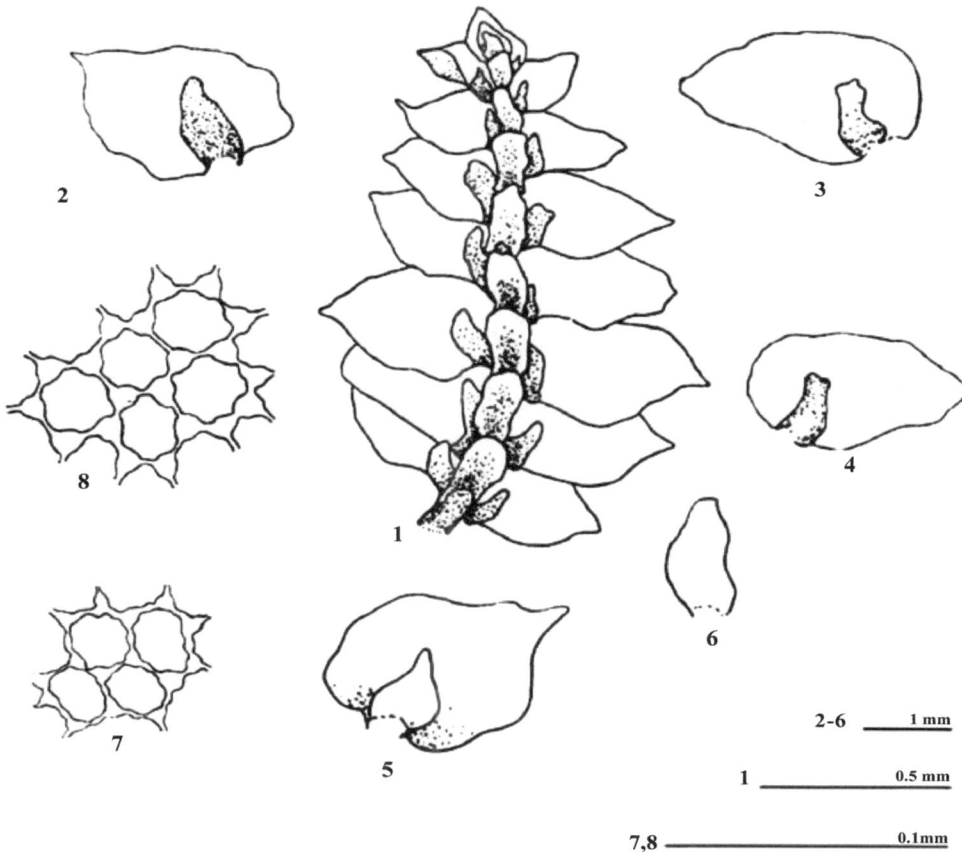

Fig. (4). 1-8: *Porella acutifolia* (Lehm. & Lindenb.) Trevis. 1. Plant, ventral view, 2-4. Leaves, 5, 6 Underleaves, 7. Basal cells of leaf –lobe, 8. Median cells of leaf-lobe (All figures drawn from LWU 12545/2000).

Plants prostrate, dark green-pale green, up to 40 mm long and 3.0-4.0 mm wide, bipinnately branched, branching '*Frullania*-type'. Leaves densely imbricate, obliquely inserted, obliquely spreading; leaf-lobes usually oblong -ovate, margins entire, 1.20-1.40 mm long, 1.19-1.28 wide, apex acute, sometime with 1-3 sub-apical teeth, cells trigonous, median cells 21-32×16-27 μm, polygonal; leaf-lobules lanceolate, entire, 0.40- 0.49 mm long, 0,.28 -0.23 mm wide, apex obtuse, slightly notched or acute, base decurrent. Under-leaves imbricate, sinuately inserted, triangular – ovate, entire, 0.48 – 0.56 mm long, 0.44 -0.048 mm wide, rarely with acuminate long decurrent base. Sterile (fertile plant not seen).

Type Locality: Annamalai hills (Tamil Nadu).

Range: Oriental region: Asia – Celebes, Hawaii, India, Indonesia (Java and Sumatra) Myanmar, Nepal, New Guinea, Philippines, Japan, Sri Lanka [13].

Distribution in India: South India: Karnataka: Kudremukh; Kerala: Devicolam; Tamil Nadu - Nilgiri hills: Ootacamund, Avalanche, Kalhatty slope, Dodabetta [13].

Ecology: Plants growing in fan like forms as epiphytic population.

Characteristics of the Species: 1. Plants small, up to 40 mm long 2. Leaf-lobes oblong-ovate, margin entire (sometimes with 1-3 sub-apical teeth) 3. Leaf-lobules slightly notched or acute, under-leaves, triangular – ovate.

Specimens Examined: South India: Tamil Nadu: Nilgiris; Ootacamund – (Dodabetta); ca. 2600 m., 08-10.2000; Legit: S.C. Srivastava and party; 12504/2000.

Porella perrottetiana is very rare in the Nilgiri hills.

Porella caespitans **(Steph.) S. Hatt. var.** *setigera* **(Steph.) S. Hatt.,** *Journ. Hattori Bot. Lab.* 33: 53 (1970); – *Madotheca setigera* Steph., *Bull. Herb. Boiss.***5**: 96 (1897); – *Madotheca calcarata* Steph., *Sp. Hep.***6**: 518 (1924**)** (Fig. **5**: 1-27 & Fig. **6**: 1-11).

Plants prostrate, dark green to light brown in color, up to 130 mm long, 2.8-3.4 mm wide, bipinnately branched, branching '*Frullania-Ptychanthus* type'. Stem 25-28 cells across the diameter, differentiated, cortical cells in 3 rows, small, 9-11×9-11 μm, thick-walled, trigonous, medullary cells large, 15-34×11-26 μm, thin-walled, trigones indistinct. Leaves densely imbricate, obliquely inserted, horizontally spreading; leaf-lobes triangular – ovate, entire, 1.6-2.88 mm long, 0.88-1.8 mm wide, apex acute-acuminate (setose), 9-10 cells long, 2-3 cells wide, up to 7 cells uniseriate; apical cells 11-26×11-15 μm, with indistinct trigones; median cells 30-45×19-26 μm, trigones nodulose; leaf-lobules large, lanceolate, entire, 0.72-1.6 mm long, 0.32-0.84 mm wide, apex obtuse to sub-acute, sometimes with one tooth, base long decurrent, sometimes recurved. Underleaves densely imbricate, transversely inserted, oblong-ovate to ovate-triangulate, margin entire, 0.84-2.04 mm long, 0.68-1.04 mm wide, apex truncate, sub-truncate or obtuse, with 0-2 (4) teeth, base longly decurrent.

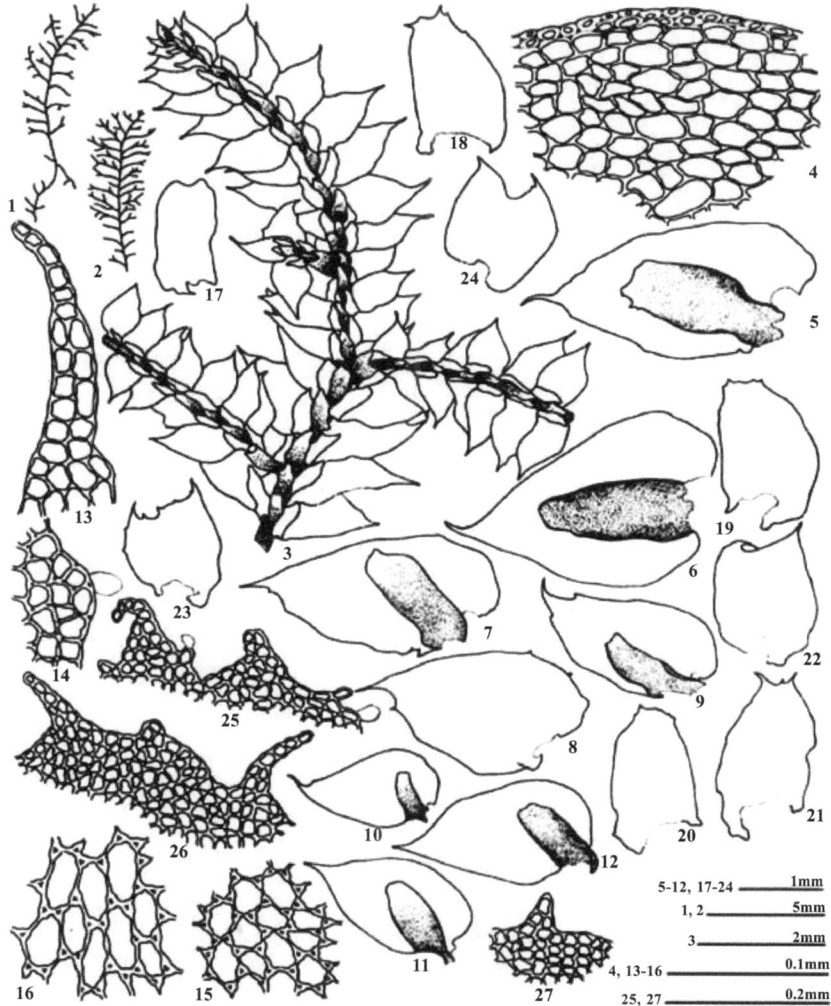

Fig. (5). 1-27: *Porella caespitans* S. Hatt. var. *setigera* (Steph.) S. Hatt. . 1, 2 Cladograph, 3. Plant, ventral view, 4. Cross section of stem, a portion, 5-12. Leaves, 13. Apical cells of leaf-lobe, 14. Marginal cells of leaf-lobe, 15. Median cells of leaf-lobe, 16. Basal cells of leaf –lobe, 17-21. Underleaves, 22-24. Branch underleaves, 25. -27. Apical cells of underleaf (All figures drawn from LWU 16971/2003).

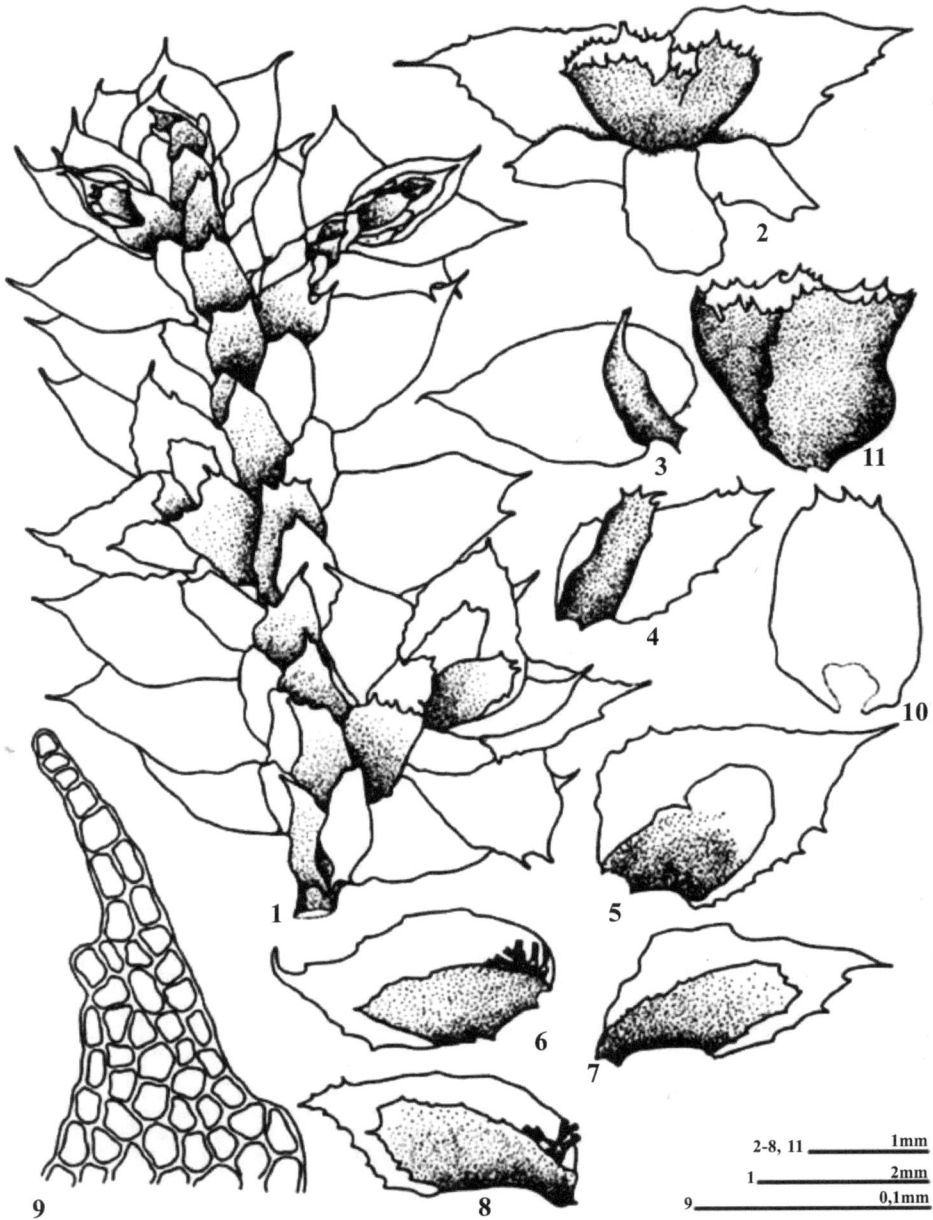

Fig. (6). 1-11: *Porella caespitans* S. Hatt. var. setigera (Steph.) S. Hatt. . 1. Femal branch, 2 Perichaetial bracts, 3-8. Female bracts, 9. Apical cells of female bracts, 10. Bracteole, 11. Perianth (All figures drawn from LWU 16971/2003).

Dioicous. Androecia not seen. Gynoecia on short lateral branches; female bracts in 1-2 pairs, 1.80-2.4 mm long, 0.88-1.5 mm wide, margins dentate; bract - lobule 0.60-1.20 mm long, 0.5-1.0 mm broad; bracteole 1.2 mm long, 0.64 mm wide, dentate. Perianth campanulate (juvenile), apex dentate with setose teeth, archegonia in clusters. Sporophyte immature.

Type Locality: Japan [16]

Range: Palearctic region: Oriental region: Asia - China, INDIA, Myanmar, Nepal, Taiwan, Vietnam. Asia- Japan, Korea [6].

Distribution in India: South India: Kerala – Vagavurrai; Tamil Nadu - Nilgiri hills [Ootacamund (Dodabetta, Glenmorgan)]; Palni hills – Kodaikanal (Shembaganur).

Ecology: Plants growing in fan life forms as epiphytic population on angiospermic trees mainly up to the height of 2 m.

Characteristics of the Species: 1. Leaf-lobe ovate-triangulate with acuminate to acute apex 2. Sub apical tooth occasionally present, apex setose 3. Underleaves oblong to ovate, triangulate, apex truncate.

Specimens Examined: South India: Tamil Nadu: Nilgiri (Dodabetta); *ca.* 2600 m; 12.02.1909; Fleischer 135; as *Madotheca calcarata* St.; Type 21504 (G). Nilgiri hills (Glenmorgan); *ca.* 2100-2200 m; 02.04.2003; P.K. Verma and A. Alam; 16971/2003, 16972/2003 (LWU).

Porella caespitans var. *setigera* was earlier known from India as *Madotheca calcarata* Steph. from Nilgiri hills and later it was synonymies by Hattori [16] under the former. The variety is very rare in distribution and has been collected only from Glenmorgan in the recent past. The plants were found with young gynoecia, while earlier workers reported only male plants. The variety is close to another taxon belonging to same species, *P. caespitans* var. *nipponica* Hatt. however, the present variety is differing in having triangular-ovate leaf-lobes with acuminate tips and sub-apical teeth (occasionally present) while *P. caespitans* var. *nipponica* is having oblong-ovate leaf-lobe with acuminate apex and frequent presence of short sub-apical teeth. Hattori [16] also separated *P. caespitans* var. *setigera* from *P. caespitans* var. *nipponica* on the basis of underleaves dentitions, which is long, 2-4 celled in the former. The plants from Nilgiri are also having varying number of underleaf dentitions ranging from obtuse, bidentate to multidentate.

Porella campylophylla (Lehm. & Lindenb.) Trevis., *In: Mem. Real Istit.*

*Lombardo Sci Lett.*ser. 3, **4**: 408 (1877); *Jungermannia campylophylla* Lehm. & Lindenb. *In*: Lehm. *Pugillus* 6: 40 (1834) (Fig. **7**, 1-10).

Plants prostrate, dark brown in colour (brown in herbarium), up to 60 mm long, 1.6-2.2 mm wide, bipinnately branched, branching '*Frullania-Ptychanthus* type'. Stem 20-22 cells across the diameter, differentiated, cortical cells in 2 (3) rows, 5-19×7-30 µm, thick-walled, trigones indistinct, medullary cells 9-43x10-36 µm, thin-walled, trigonous. Leaves densely imbricate, widely spreading, obliquely inserted; leaf-lobe ovate, oblong-ovate, 1.4-2.1 mm long, 0.75-0.94 mm wide, apex sub-truncate, rounded, obtuse, with 3-7 sharp teeth, teeth blunt, 6-7 cells in height, 2-3 cells wide; apical cells 19-32×15-23 µm, trigones indistinct, median cells (19) 30-48×16-37 µm, rounded, trigonous, basal cells 32-50×25-34 µm, hexagonal, trigonous; leaf –lobule lanceolate, 0.60-1.14.mm long, 0.15-0.20mm wide, apex obtuse (acute), sometimes slightly notched with 1-2 small teeth, base long decurrent. Underleaves closely imbricate, transversely inserted, oblong-ovate, 0.78-1.16mm long, 0.42-0.82 mm wide, apex obtuse, truncate (acute) with (0) 1-4 small teeth, base longly decurrent (sometimes crispate). Plants sterile.

Type Locality: Nepal [6].

Range: Oriental region: Asia- Bhutan, Cambodia, China, India Laos, Thailand; Vietnam, Myanmar, Nepal [6].

Distribution in India: Eastern Himalayas: Arunachal Pradesh – Kameng; Manipur: West Mountain; Meghalaya – Cherrapunji, Mawphlang, Manebhanjang, Pynurshala; Sikkim; West Bengal – Kurseong, Tongloo, Jorpokhri, Mungpoo, Darjeeling. Western Himalayas: Himachal Pradesh: Great Himalayan- National park; Uttarakhand – Mussoorie, Nainital, Pindari, Ranikhet. South India - Tamil Nadu – Nilgiri hills [Ootacamund (Glenmorgan, Government Botanical Garden), Upper Bhavani (Avalanche)]; Palni hills- Kodaikanal (Shembaganur) [13, 15, 20].

Ecology: Plants growing in fan like life forms as the epiphytic population on big trees.

Characteristics of the Species: 1. Leaf-lobe ovate-oblong, apex sub-truncate and obtuse with 3-7 teeth 2. Leaf-lobule lanceolate-ligulate, longly decurrent 3. Under-leaves oblong-ovate, apex obtuse, base longly decurrent.

Specimens Examined: Nepal: *Madotheca campylophylla* Lehm. & Lindenb.; Nepal; (Ex herb. Rome); G 21507 (TYPE).

India: South India: India Orientalis: As *Madotheca indica* Steph.; Pfliderer; G 21527.Tamil Nadu: As *Madotheca madurensis* Steph.; Palni hills – Kodaikanal

(Shembaganur-Madura); 1911; R. P. G. Foreau; G 16796. Nilgiri hills –
Ootacamund (Glenmorgan); *ca.* 2100-2200 m; 01.12.2001; P.K. Verma and A.
Alam; 14699/2001 (LWU). On way to Glenmorgan; i. 2133 m.; 01.10.2002; P.K.
Verma and A. Alam; 16256/2002, 16261/2002, 16266/2002 (LWU).

Porella campylophylla is the sole species of the *campylophylla* complex. This
species is represented with 3 variety *P. campylophylla* var. *ligulifera*, *P.
campylophylla* var. *ptychantha* and *P. campylophylla* var. *lancistipula* (Shaheen
and Srivastava, 1989). Out of which only *P. campylophylla* var. *ligulifera* has
been reported from Nilgiri hills. Two species of *Porella*, *P. indica* Steph. and *P.
madurensis* Steph. earlier known from south India now come under *P.
campylophylla* [6].

This species is very rare in Nilgiri hills, and recently found growing in Avalanche
and Glenmorgan only.

Porella campylophylla var. ligulifera (Taylor) S. Hatt., *Journ. Hattori Bot.
Lab.***32**: 333 (1969); Hattori, *Ibid***44**: 101 (1978). – *Madotheca ligulifera* Taylor,
in Lehm. *Pugills*. 10 (1844) (Fig. **8**: 1-32).

Plants prostrate, in compact patches, brownish-green (chestnut) in color, up to 110
mm long, 5 mm wide, irregularly bipinnately branched, branching usually
'*Frullania-Ptychanthus* type'. Stem 20-30 cells across the diameter,
differentiated, cortical cells in 3-4 rows, small, 13-15×15-19 µm, thick-walled,
medullary cells large, 26-30×19-28 µm, thin-walled, trigonous. Leaves densely
imbricate, obliquely inserted; leaf-lobe triangular-ovate, 1.1-2.5 mm long, 0.8-1.1
mm wide, margins with 3-11 sharp teeth, teeth 5-8 cells long, 3-4 cells wide, apex
acute to acuminate, ventral margin strongly recurved and arched; apical cells20-
30×15-22µm, trigones small, median cells (24) 34-50×17-40 µm, trigones large,
basal cells large, elongated, 38-58×24-32 µm; leaf-lobule lanceolate, canaliculate,
0.66-1.22 mm long, 0.17-0.24 mm wide, apex obtuse, often with 1-4 small teeth,
lateral margin incurved, base longly decurrent. Underleaves oblong-ovate to
triangular, 0.88-1.28 mm long, 0.48-0.80 mm wide, apex obtuse to truncate, with
2-11 acute teeth, lateral margin incurved, base longly decurrent. Plant sterile.

Type Locality: Nepal [6].

Range: Oriental region: Asia- INDIA, Nepal, Thailand [6].

Distribution in India: Eastern Himalayas: Arunachal Pradesh – Kameng;
Manipur, Senapati; Meghalaya –Cherrapunji, Jowai, Manebhanjang, Mawphlang;
Sikkim – Gangtok. Western Himalayas: Uttaranchal – Mussoorie. South India:
Tamil Nadu - Nilgiri hills [Upper Bhavani (Avalanche)] [13, 20, 22 - 24].

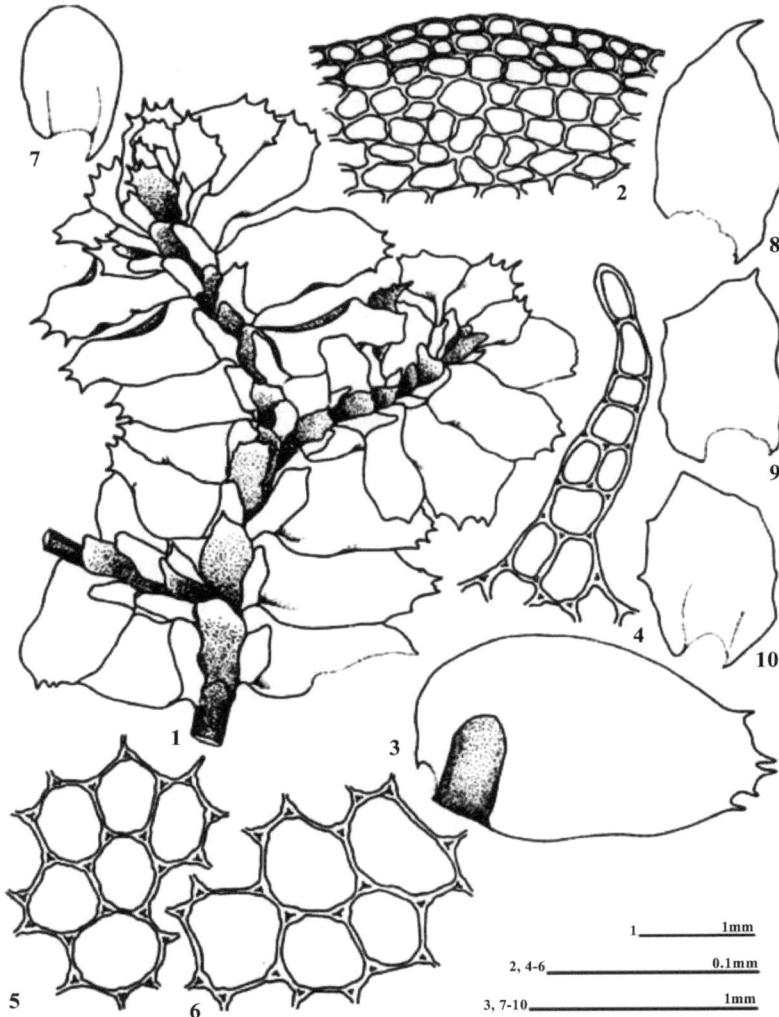

Fig. (7). 1-10: *Porella campylophylla* (Lehm. & Lindenb.) Trevis. 1. Plant, ventral view, 2. Cross section of stem, a portion, 3. Leaf, 4. Apical cells of leaf-lobe, 5. Median cells of leaf-lobe, 6. Basal cells of leaf –lobe, 7-10. Underleaves (All figures drawn from LWU 14699/2001).

Ecology: Plants growing in fan like life forma as epiphytic population on decaying logs and basal portion of angiospermic trees.

Characteristics of the Species: 1. Leaf-lobes triangular-acute with acute or acuminate tips 2. Ventral margin of leaf-lobe strongly recurved 3. Underleaves large and strongly acute to strongly toothed at apex.

<u>***Specimen Examined:***</u> South India: Tamil Nadu: Nilgiri hills- Upper Bhavani (Avalanche); *ca.* 2250 m; 09.10.2000; S.C. Srivastava and party; *12523/2000, 12534/2000, 12553/2000, 12558/2000, 12561/2000, 12562/2000, 12564/2000, 12565/2000, 12568/2000, 12573/2000, 12574/2000, 12575/2000, 12584/2000, 12586/2000, 12587/2000, 12593/2000* (LWU).

This variety was reported from India by Mitten [9] from the Nilgiri hills as *Madotheca ligulifera* Taylor, while Hattori [25] reported taxa from eastern and western Himalayas. Recently this variety has been recollected from Avalanche (Nilgiri hills).

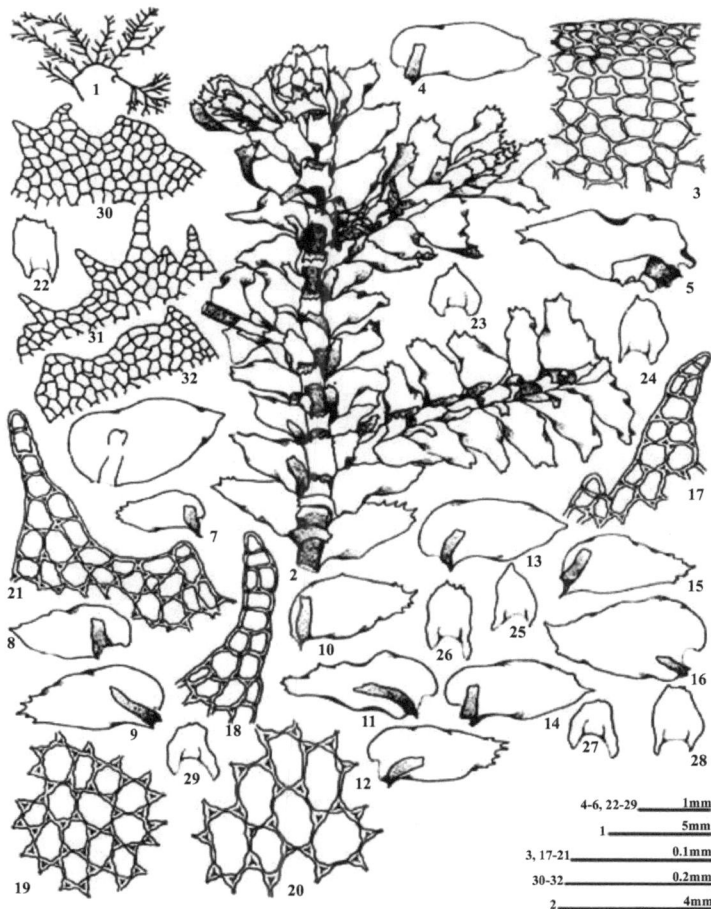

Fig. (8). 1-32: *Porella campylophylla* (Lehm. & Lindenb.) Trevis. var. *ligulifera* (Tayl.) S. Hatt. 1. Cladograph, 2. Plant, ventral view, 3. Cross section of stem, a portion, 4-16. Leaves, 17, 18, 21. Apical cells of leaf-lobe, 19. Marginal cells of leaf-lobe, 20. Basal cells of leaf-lobe, 22-29. Underleaves, 30-32. Apical cells of underleaf (All figures drawn from LWU 12573/2000).

This species is closely related to sole species of *campylophylla* complex '*P. campylophylla*' in wider, ovate to ovate-oblong leaf-lobe, but differs in number of dentitions in leaf-lobe (up to 11 in *P. campylophylla* var. *ligulifera* and 3-7 in *P. campylophylla*), recurved margin of leaf-lobe in *P. campylophylla* var. *ligulifera*) and underleaves (obtuse with 1-4 dentition in latter, while truncate with up to 11 teeth in the former).

CONSENT FOR PUBLICATION

Not applicable.

CONFLICT OF INTEREST

The authors confirm that this chapter contents have no conflict of interest.

ACKNOWLEDGEMENTS

The author is grateful to Professor S. C. Srivastava, former Head of the Botany Department, Lucknow University, Lucknow for facilities and encouragement, the Ministry of Environment, Forests and Climate Change, New Delhi for financial assistance under AICOPTAX, and the Department of Forest and Environment, Tamil Nadu for permission and logistic support for the period of field explorations.

REFERENCES

[1]   Schuster RM. The Hepaticae and Anthocerotae of North America East of the Hundredth Meridian. New York: Columbia University Press 1980; Vol. IV.

[2]   Verma PK, Alam A, Srivastava SC. Status of mosses in Nilgiri hills (Western Ghats), India. Arch Bryol 2011; 102: 1-16.

[3]   Verma PK, Alam A, Rawat KK. Assessment of liverwort and hornwort flora of Nilgiri hills, Western Ghats (India). Polish J Bot 2013; 58(2): 525-37.
[http://dx.doi.org/10.2478/pbj-2013-0038]

[4]   Steere WC, Schuster RM. The Hepatic genus *Ascidiota* Massalongo new to North America. Bull Torrey Bot Club 1960; 87(3): 209-15.
[http://dx.doi.org/10.2307/2482767]

[5]   Qian K, Bi XF, Shu L, Zhu RL. Porella longifolia (Steph.) S.Hatt. and Porella densifolia var. robusta (Steph.) S.Hatt. (Porellaceae, Marchantiophyta) excluded from the liverwort flora of China. Phytotaxa 350(2): 182.
[http://dx.doi.org/10.11646/phytotaxa.350.2.9]

[6]   Hattori S. Studies on the Asiatic species of the genus *Porella* (Hepaticae). VII. A Synopsis Porellaceae. J Hattori Bot Lab 1978; 44: 91-120.

[7]   Mufeed B, Manju CN. *Porella perrottetiana* (Porellaceae, Marchantiophyta) a species from the Western Ghats of Kerala. Acta Bot Hung 2017; 59(1–2): 269-72.
[http://dx.doi.org/10.1556/034.59.2017.1-2.9]

[8]   Montgne C. Cryptogamae nilgherienses seu plantarum cellularium in montibous penensulal indicae

Neel-gherries dictis in a d Perrottet collectarum enumeration. Ann Sci Nat II 1842; 17: 12-23.

[9]   Mitten W. Hepaticae Indiae Orientalis: an enumeration of the Hepaticae of the East-Indies. J Proc Linn Soc Bot 1861; 5: 385-92.
[http://dx.doi.org/10.1111/j.1095-8312.1861.tb01343.x]

[10]   Stephani F. Species Hepaticarum. Geneve 1917 – 1924; 6: pp. 433-763.

[11]   Chopra RS. Notes on Indian Hepaticae. I. South India. Proc Indian Acad Sci ser B 1938; 7: 239-51.

[12]   Hattori S. Studies of the Asiatic species of the genus *Porella* (Hepaticae). I. J Hattori Bot Lab 1967; 30: 129-51.

[13]   Shaheen F. Studies in Indian Porellaceae (A monographic study) 1983.

[14]   Hattori S. Studies on the Asiatic species of the genus Porella (Hepaticae). V. J Hattori Bot Lab 1975; 39: 269-76.

[15]   Singh SK, Singh DK. Contribution to the bryoflora of Great Himalayan National Park, Kullu, Himachal Pradesh IV: Genus Porella (Porellaceae). Geophytology 2006; 36(1&2): 93-107.

[16]   Hattori S. Studies of the Asiatic species of the genus *Porella* (Hepaticae). III. J Hattori Bot Lab 1970; 33: 41-87.

[17]   Hong SW. The Hepaticae and Anthocerotae of the Korean peninsula: identification keys to the taxa. Lindbergia 2003; 28: 134-47.

[18]   Yamada K, Choe DM. A checklist of Hepaticae and Anthocerotae in the Korean Peninsula. J Hattori Bot Lab 1997; 81: 281-306.

[19]   Jones EW. African Hepatics. SXVI. *Porella* in Tropical Africa. Trans Brit Bryo Soc 1963; 4(3): 446-61.
[http://dx.doi.org/10.1179/006813863804812327]

[20]   Singh D, Dey M, Singh DK. A synoptic flora of liverworts and hornworts of Manipur. Nelumbo 2010; 52: 9-52.

[21]   Udar R, Shaheen F. *Porella hattorii* sp. nov. from India. Lindbergia 1983; 9: 70-2.

[22]   Parihar NS, Lal B, Katiyar N. Hepatics and Anthocerotes of India A New annotated checklist. Allahabad: Central Book Depot 1994.

[23]   Shaheen F, Srivastava SC. *Porella campylophylla* (Lehm. *et* Lindenb.) Trev. complex in India. Geophytology 1989; 19(1): 34-8.

[24]   Udar R, Shaheen F. Morpho-taxonomy of *Porella perrottetiana* (Mont.) Trev. from South India. J Indian Bot Soc 1983; 62: 319-25.

[25]   Hattori S. Studies on the Asiatic species ofthe genus *Porella* (Hepaticae) II. J Hattori Bot Lab 1969; 32: 319-59.

CHAPTER 10

Habitat Range of Bryophytes: A Pictorial Representation

Shiv Charan Sharma[1], Abhishek Tripathi[2], Krishna Kumar Rawat[3], Sonu Yadav[4] and Afroz Alam[1,*]

[1] *Department of Bioscience and Biotechnology, Banasthali Vidyapith, Rajasthan- 304022, India*

[2] *PCS, Government of Uttarakhand, Uttarakhand India*

[3] *CSIR-National Botanical Research Institute, Lucknow, India*

[4] *Department of Botany, University of Lucknow, India*

Abstract: The habitat and organism interaction decide the fate of the organism regarding its survival and expansion. Every organism has its own preferred habitat to reside and reproduce. Habitat loss is one of the major reasons of species extinction and endemism. A very specific relationship exists between these two components of ecosystem where rigid species become extinct and flexible prevail. In this study, habitat range of bryophytes from two bryological regions has been presented in a pictorial way to amplify the perspective regarding the habitat flexibility of the first land plants.

Keywords: Bryophytes, Corticolous, Epiphytes, Habitats, Terricolous.

INTRODUCTION

Bryophytes have an interesting place within the Plant Kingdom and it has been considered that they should have their individual sub-kingdom because of the uniqueness of the members of Anthocerotopsida. These hornworts have small size and self-sufficient, prevailing gametophyte and reliant sporophyte with the Bryopsida and Hepaticopsida, have been well thought-out by most taxonomists now to be in a new phylum (=division), the Anthocerotophyta [1]. Likewise, bryologists also agree that the liverworts should have a separate phylum as Marchantiophyta leaving the only the members of the bryophyta (Musci). Collectively, the mosses, liverworts, and hornworts are still considered as bryophytes in its common sense.

The gametophytes of bryophytes are among the most complex forms among the

---

[*] **Corresponding author Afroz Alam:** Bryotechnology Laboratory, Department of Bioscience and Biotechnology, Banasthali Vidyapith, Rajasthan, India; Tel: +91-9415596994; E-mail: afrozalamsafvi@gmail.com

plants [2]. Bryophytes appear more intricate because of their diminutive size. Some bryophytes are only a few millimetres high and have very few leaves [3]. *Buxbaumia* is well-known which has a hefty capsule on a bulky stalk, but only a small number of unique leaves for the protection of the archegonia; and depends initially on its protonema, and later on the capsule for food material. Likewise, *Monocarpus* (liverwort) thallus is about 0.5-2.0 mm in diameter. In contrast, *Polytrichum commune* (moss) has well developed foliage and scales [3] while, *Fontinalis* spp. (aquatic moss) can grow up to 2 m span [4].

Usually, a view of lush and green tropical scene appears when we think about the natural habitats of the bryophytes. However, they are almost all-pervading and a bryologist can collect these amphibian of the plant kingdom from the freezing arctic and antarctic to the scorching deserts of the globe and from sea level to gigantic alpine peaks except ice. Likewise, various species of bryophytes can grow on sand and in deserts except on the extremely movable sand dunes in desolate tract (Plate 1: Figs. **1-6**). For instance, the thallose liverwort *Monocarpus sphaerocarpus* is reported in quite harsh habitats like saline or gypsum-rich soils of Australia and Africa [4 - 8].

In this chapter a comparison is made amid the natural bryo-habitats of Uttarakhand (Himalayan region) and Rajasthan (Desert region) of India (Plate 2: Figs. **1-6**) in a pictorial way. This study aims to provide information regarding the occurrence of these interesting plants in both the regions of India. An exhaustive account in the form of pictures is given which through a light on the preferred habitats of bryophytes, according to the contrasting environmental conditions of both the regions.

This will also demonstrate various adaptations according to prevailing challenges by the bryophytes in entirely different climatic conditions. This chapter gives a pictorial information about the substrate on which these plants prefer to grow.

There are diverse factors that persuade the spread of bryophytes, *viz.*, climate (especially annual temperature and the patterns of rainfall); chemical constitution of the substrate (*e.g.,* alkaline/acidic); physical factors *viz.*, texture of surface, extent of shade; and the level of pollution in the ambient environment.

Bryophytes dwell in diverse microhabitats. Some arise absolutely in the moist, shady under-storey of slightest disturbed forests, and in the interior tree canopy high above the land. Some prefer river banks and soggy road-side cuts to colonize. Nevertheless, generally they occur on soil and muddy walls (terricolous), pebbles and stones (saxicolous), rocks (rupicolous) and decaying logs (lignicolous). Besides, growing as epiphytes on tree trunks (corticolous) and leaves (epiphyllous/folicolous) are of frequent incidence (Plate 4: Figs. **1-6**). A distinguished feature of some bryophytes is their amazing capability to tolerate drought and are usually known as the xeromorphic forms. They form the secondary colonizers after lichens on barren rocks (Plate 5: Figs. **1-6**).

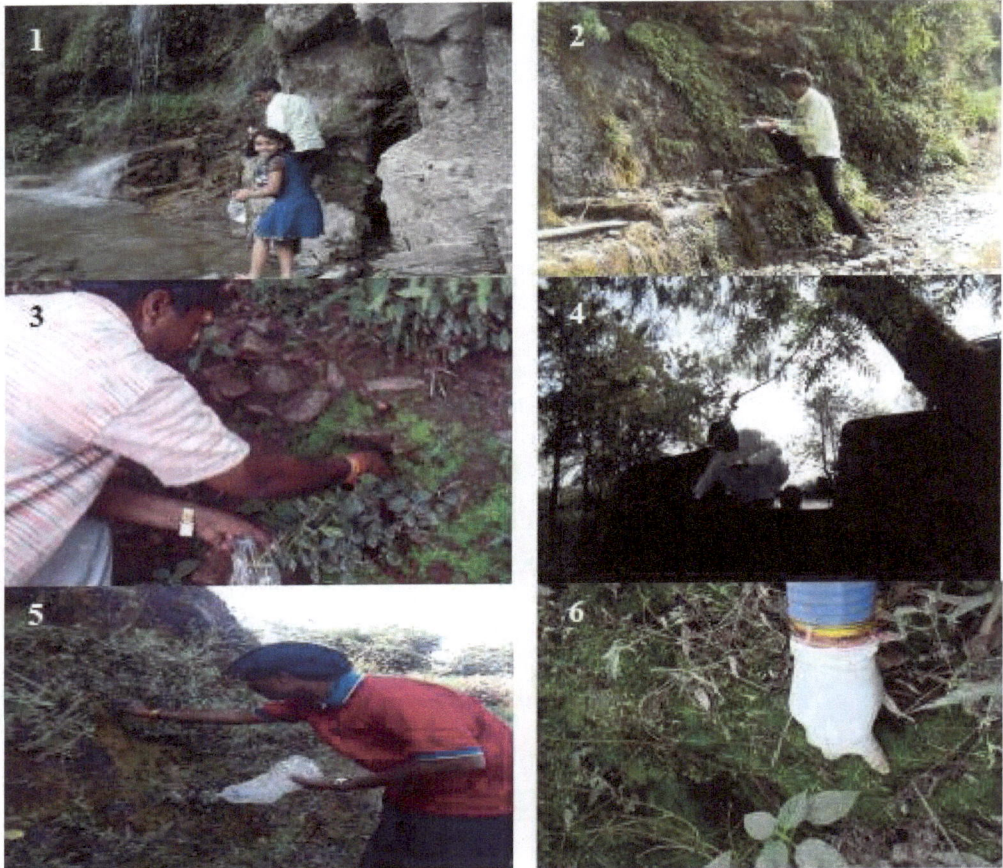

Plate 1. Figs: **1-6**: Collection of bryophytes from **1**. Water source, **2**. Exposed rocks, **3**. Wet soil, **4**. Old wall, **5**. Shady rock, **6**. Moist ground.

There are some taxa that are limited to very exact substrate/habitats, but most of the bryophytes exist in diverse habitats/substrate. In the few instances that's not too unexpected. For instance, a completely rotted timber need not be much dissimilar to affluent, organic top soil in terms of quality, chemical properties, and water-holding competence. It would then expect to find a few species growing on both well-rotted wood and soil. There are some species of bryophytes which are very specific to alkaline substrate, others prefer acidic substrate and some can grow on both the substrate. The Saxicolous bryophytes that need alkalinity could be collected on limestone, but by no means on basalt (acidic rock). A bryologist could also discover only on the alkaline soil. Same species can be discovered on the surface of brick/concrete wall/roofs because the alkaline nature of concrete. In case of clay walls lime-preferring bryophytes (Plate 7: Figs. **1-6**). can be observed. In some cases rotted old bones of animals are preferred by lime-loving species of bryophytes [4 - 8].

Plate 2. Figs: **1-6**: Collection of bryophytes from **1**. Tree bark, **2**. Tree base, **3**. Wet rocks, **4**. Road side slopes, **5**. Moist and shady rocks, **6**. Shaded ground.

Extensive types of habitats, for instance, grasslands, rainforests and deserts are momentous, but it is also very imperative to come across at the small scale when it concerns to bryophytes. Because, inside a solitary all-embracing habitat there are always abundant micro-habitats. The variety of surfaces for instance, boulders, logs, depressions, trees– all occur in an area of study creating a different float up, light exposure, chemical contents, water flow patterns, *etc.* All are creating micro-habitats which provide habitats to diverse bryophytes (Plate 4: Figs. **1-6**).

Plate 3. Figs: **1-6**: Terrestrial Bryophytes: **1**. *Riccia gangatica*, **2**. *Anthoceros crispulus*, **3**. *Cyathodium cavernarum*, **4**. *Plagiochasma rupestre*, **5**. *Marchantia polymorpha*, **6**. A mixed population of *Anthoceros* and mosses.

Plate 4. Figs: **1-6**: Terrestrial Bryophytes: **1**. Colonization on burnt land by mosses, **2**. Thalloid liverworts on wet soil, **3**. Growth of moss on rocky substratum, **4**. Leafy liverworts on sun facing tree trunk, **5**. Patch of wet soil with mixed population of liverworts and mosses, **6**. A mixed population of leafy, thalloid liverworts and mosses.

Great diversity of habitats exerts substantial confronts to routine continued existence consequently, bryophytes have adapted a range of strategies to conquer these challenges. In the desert, arctic and antarctic regions, an understandable confront is water, which is characteristically scanty and random – however, there are numerous bryophytes in these arid areas, with different endurance strategies.

One approach to conquer the challenge is to nurture in sheltered areas, such as, in crevices of rock or at the boulders' base or basal part (Plate 2: Fig. **2**) of the shrubs in that way receiving a little defence against intense winds. Many of the desert species that exist in more exposed areas raise as crowded colonies in the form of a carpet or cushion, which facilitates to guard against speedy winds and also helps preserve water (Plate 5: Figs. **3**, **4**). Incidentally, the tree trunk is also considered as an arid environment, if it is bare and receiving direct sun and speedy winds. In such situations, the bryophytes are likely to grow in the cracks of the bark as small and crowded cushions. In urban areas this mode of growth is easy to observe where both opportunities and challenges occur concurrently.

Plate 5. Figs: **1-6**: Terrestrial Bryophytes: **1**. Colonization of semi open substratum by *Plagiochasma* spp., **2**. Thalloid liverworts with elevated reproductive structure for utmost dispersal, **3**. Growth of moss on fallen parts of trees, **4**. A cushion of moss growth on moist substratum, **5**. A patch of mosses showing the rhizoidal parts, **6**. A population of moss with dependent sporophyte.

A disturbed land exerts its own challenges. It is evident that many gametophytes in such a land area would be hidden or broken during the regular roadside cutting, ploughing, with comparatively few surviving unbroken from one year to another. Effective methods for inhabiting such habitats are the quick maturity of the spores or vegetative propagules or the assembly of resting bodies such as tubers/gemmae (Plate 6: Figs. **1-6**).

Plate 6. Figs: **1-6**: Terrestrial Bryophytes: **1**. Colonization of wet substratum by *Marchantia* spp., **2**. Thalloid liverworts *Plagiochasma rupestre* on exposed rocks, **3**. Growth of *Asterella* spp. on steep rock, **4**. Growth of *Asterella* spp. with mosses, **5**. A patch of mixed population of sun loving bryophytes, **6**. A mat of moss on exposed rock.

Plate 7. Figs: **1-6**: Terrestrial Bryophytes: **1**. Colonization of calcium rich substratum by thalloid liverworts., **2**. Growth of moss as mat on semi-exposed soil, **3**. Growth of moss on completely exposed rock, **4**. Growth of mosses in cervices of rocks, **5**. A patch of thalloid liverwort with mosses on semi-wet rocks **6**. A mat of moss on damp soil.

The thriving bryophyte colonies at the margins of water bodies like streams or lakes, on boulders in streams and even on the rock faces of water cascades. Few of them can also bear drier habitats while others cannot stay alive away from an aquatic surroundings. The bryophytes in these aquatic sites are recurrently speckled with water and with time may even become submerged for comparatively short periods. However, there are also eternally submerged species of bryophytes while few species can glide on the water's surface [9]. Overall, a water body provides a variety of micro-habitats for these bryophytes. Similarly, bogs also dwell to many bryophytes, *e.g., Sphagnum* (the bog moss).

In the desert, sand is both a substrate and a confront to many terrestrial species of bryophytes (Plate 3: Figs. **1-6**), which, in these areas may simply be covered by windblown sand grains. Nevertheless, bryophytes do raise on sand, even on the sand dunes excluding those that are extremely movable. The moss *Tortella* sp. is the best known example emerging with other higher plants on an undersized, near the coastal sand dune [4 - 8].

DISCUSSION

Every living organism lives in a habitat, according to micro and macro climatic conditions, the organisms that live within it must either adapt or escape out in order to survive. The latter option is usually preferred by fauna while flora need special strategies to tolerate or adapt the environmental conditions due to the lack of escaping organs. Flexibility in habitats confers the survival and dispersal of organisms. Bryophytes can be seen in immense array all through the evolution in regions varying from arid/semiarid to tropical rainforest, and in locale from sea-level to alpine. They crop up most copiously in somewhat unpolluted regions. However, few exceptions are there as some species prefer selective habitat, but most of the bryophytes are found in a range of habitats. The present study reveals the huge range of habitats from high altitude to desert conditions. However, the diversity, frequency and abundance may vary in accordance to the environmental impact. They are considered as first land plant and pioneers along with lichens because of their extraordinary ability to adapt the various climatic conditions. Bryophytes are generally neglected due to their small size in the recent past, but now with the ever increasing demand of novel genome and phytochemical these plants should be focused more because they have an amazing reservoir of nature's mystery and offerings.

CONSENT FOR PUBLICATION

Not applicable.

CONFLICT OF INTEREST

The authors confirm that this chapter contents have no conflict of interest.

ACKNOWLEDGEMENTS

The authors are grateful to Prof. Aditya Shastri, Vice Chancellor, Banasthali Vidyapith (Rajasthan), India and Governments of Uttarakhand and Rajasthan for their help and support.

REFERENCES

[1]   Shaw J, Renzaglia K. Phylogeny and diversification of bryophytes Amer J Bot 2004; 91: 1557-81.
[http://dx.doi.org/10.3732/ajb.91.10.1557]

[2]   Renzaglia KS, Nickrent DL, Garbary DJ, Garbary DJ. Duff RJT. Vegetative and reproductive innovations of early land plants: implications for a unified phylogeny. Philos Trans R Soc Lond B Biol Sci 2000; 355(1398): 769-93.
[http://dx.doi.org/10.1098/rstb.2000.0615] [PMID: 10905609]

[3]   Crum HA. Structural diversity of bryophytes. Ann Arbor, MI: The University of Michigan Herbarium 2001.

[4]   Glime JM. Meet the Bryophytes In: Glime JM, Ed. Bryophyte Ecology. Physiological Ecology, Ebook 2-1-1 Michigan Technological University and the International Association of Bryologists 2017; 1.

[5]   Porley R, Hodgetts N. Mosses and Liverworts. London: Collins 2006.

[6]   Meagher D, Fuhrer B. A Field Guide to the Mosses and Allied Plants of Southern Australia. Canberra, Melbourne: Australian Biological Resources Study & Field Naturalists Club of Victoria 2003.

[7]   Ramsay HP. Introduction to Mosses Flora of Australia. 2006; 51 (Mosses 1).

[8]   Scott GAM, Stone IG. The Mosses of Southern Australia. London: Academic Press 1976.

[9]   Alam A. Textbook of Bryophyta. New Delhi, India: I K International Publishers 2015.

CHAPTER 11

Bryodiversity, Threats and Conservation of Liverworts and Hornworts of Kolhapur District (Maharashtra), India

Rajendra Ananda Lavate[*]

Raje Ramrao Mahavidhyalaya, Jath-416 404; Dist. Sangli (M.S.), India

Abstract: Bryodiversity, including liverworts and hornworts strongly concentrated in the humid tropics is ecologically very significant. The present paper provides the first hand consolidated account of 42 species of liverworts and hornworts belonging to 19 genera and 13 families of Kolhapur District. It includes a detailed account of the bryodiversity, causes of threats and action plan for their conservation, monitoring and management.

Keywords: Bryodiversity, Liverworts, Hornworts, Conservation, Maharashtra.

INTRODUCTION

Liverworts and hornworts are small terrestrial plants that grow in various life forms *viz.*, mats and cushions found in all habitats and climatic conditions except marine environment. They flourish well in the moist mountain forests of the tropics and subtropics as well as in Arctic, Tundra and Antarctic regions. Usually, they are most common in rainy and humid areas. However, they give a preference for microclimate niches such as moist, shaded ground, moist rocks, bogs, rock crevices, barks and the vicinity of small shady springs where the microclimate is often different from conditions recorded by standard meteorological methods.

In India, studies on liverworts and hornworts, explored, evaluated and elaborately reviewed from time to time by many bryologists of India [1 - 28]. According to Singh *et al.* [26], there are about 930 species of liverworts and hornworts that belong to 140 genera and 59 families, out of which 834 species, 16 subspecies, 39 verities and 2 forma, belong to 134 genera, 56 families of liverworts and 39 species belong to 6 genera in 3 families of hornworts.

---

[*] **Correspondence Rajendra Ananda Lavate:** Raje Ramrao Mahavidhyalaya, Jath-416 404; Dist. Sangli (M.S.), India; Tel: +91-9623420151; E-mail: bryoraj@gmail.com

Afroz Alam (Ed.)

Amazingly, as far as Maharashtra is concerned, studies on liverworts and hornworts are very meagre. In the past, an attempt was made by Morajkar [29] who studied liverworts from Nasik. Joshi and Biradar [6] have given distribution and enumeration of liverworts from Western Ghats of Maharashtra. Kalgaonkar [30] and Barve [31] have given monographs and histochemical accounts of some hepatic taxa and certain thalloid liverworts, respectively from Maharashtra. Chavan [32], Apte and Sane [33], Gupte [34], Dabhade [35], Joshi [36, 37], Shirke [38] have worked on bryophyte flora. Dabhade [39] reported 6 species of *Riccia* (Mitch.) L. from Maharashtra *viz.*, *R. discolor*, *R. billardieri*, *R. plana*, *R. fluitans*, *R. frostii* and *R. indiragandhii*. The last one *R. indiragandhii* was the new addition to the region by Dabhade and Hasan [40]. Chaudhary *et al.* [17] described 100 species of bryophytes from North Konkan of Maharashtra. Out of these 23 species are of liverworts, 18species are of hornworts and 59 are of mosses. Recently, Bagawan and Kore [41] have studied the distribution of liverworts and hornworts of Kas plateau and reported 20 species of liverworts and hornworts belonging to 12 genera, 8 families and 3 orders.

Currently, Singh *et al.* [26] reported about 72 species of liverworts and hornworts belong to 25 genera and 18 families of which 52 species are of liverworts, 20 species are of hornworts from Maharashtra.

Bryodiversity and Status of Kolhapur District

Kolhapur is the southernmost district of Maharashtra state having geographical area of about 7685 sq km located between 17° 17′ to 15° 43′ North latitudes and 73° 40′ to 74° 42′ East longitudes at an average elevation from 390 to 900 meters entirely in the Panchganga basin. It includes the main range of Sahyadri running North-South on Western Side and large spurs which stretch North-East and East from Sahyadris and valleys with about 16 forts of historical and botanical interest as well, *viz.*, Aajra, Bhudargad, Chandgad, Gagangad, Gandharvgad, Hanumantgad, Kalanidhigad, Mahipaalgad, Paargad, Paavangad, Panhala, Saamangad, Shivgad, Vallabhgad, Vilasgad and Vishalgad [25]. Lavate [42] has studied liverworts of Panhala. Dongare [43] has given an ecological assessment of the liverworts of Panhala Hill station. Lavate *et al.* [44, 45] reported *Marchantia linearis* and *Pallavicinia lyelli* from Kolhapur District, respectively which are new additions to the hepatic flora of Maharashtra. Lavate [28] has given detailed account of bryodiversity, distribution, threats and conservation status of 41 species of liverworts and hornworts belonging to 18 genera and 12 families from forts of Kolhapur District.

The present comprehensive survey of Kolhapur District records 42 taxa belonging to 19 genera of liverworts and hornworts, belonging to 13 families of 4 orders

(Fig. **1**). These 42 taxa (32 liverworts and 10 hornworts) belonging to 3 orders, 11 families and 15 genera of Hepaticopsida and 1 order, 2 families and 4 genera of Anthocerotopsida in the Kolhapur District of Maharashtra (Table **1** and **2**). This accounts for 4.1 percent of total Indian liverworts and hornworts. Out of 42 taxa, 15 genera and 32 species were belonging to the group Hepaticopsida while 4 genera and 10 species were belonging to the group Anthocerotopsida. Amongst these 7 species were reported as new distributional records for Maharashtra.

List of Genera and Species

I] Order: Jungermanniales

  i) Family: Jungermanniaceae
  Solenostoma **Mitt.**
    1) *S. fossomroniodes* Kashyap
    2) *S. tetragonum* (Lindenb.) R.M. Schust.
  ii) Family: Lejeuneaceae Casaers-Gil
  Archilejeunea **(Spruce) Schiffn.**
    3) *A. minutilobula* Udar *et* Awasthi
  Cheilolejeunea **(Spruce) Schiffn.**
    4) *C. intertexta* (Lindenb.) Steph.
  Lejeunea **Libert.**
    5) *L. discreta* Lindenb.
    6) *L. flava* (Sw.) Nees
    7) *L. tuberculosa* Steph.

II] Order: Metzgeriales
  iii) Family: Aneuraceae
  Aneura **Dumort.**
    8) *A. pinguis* (L.) Dumort.
  Riccardia **Nat.**
    9) *R. levieri* Schiffn.
    10) *R. santapaui* Udar and S. C. Srivast.
  iv) Family: Fossombroniaceae *Fossombronia*
    11) *F. himalayensis* Kash.
    12) *F.indica* Steph.
  v) Family: Metzgeriaceae *Metzgeria* **Raddi**
    13) *M. himalayensis* Kash.
  vi) Family: Pallaviciniaceae *Pallavicinia* **Gray**
    14) *P. lyellii (Hook.) Gray.*

III] Order: Marchantiales
  vii) Family: Aytoniaceae Asterella P. Beauv.

15) *A.wallichiana (Lehm.Et Lindenb.) Grolle*

=*Plagiochasma* **Lehm.and Lindeb.**

16) *P. appendiculatum* Lehm.and Lindeb.

17) *P. intermedium* Lindenb. and Gottsche

18) *P. pterospermum* C. Massal.

viii) Family: Cyathodiaceae *Cyathodium*

19) *C. cavernarum* Kunze, Lehm.

20) *C. epiphytens Bira.*

21) *C. tuberosum* Kash.

ix) Family: Marchantiaceae *Marchantia*

22) *M. linearis* Lehm.and Lindenb.

x) Family: Targioniaceae *Targionia*

23) *T. hypophylla* L.

24) *T. hypophylla* (Mich.) L.var. *sinhagarhii* Kalgaonkar

xi) Family: Ricciaceae *Riccia*

25) *R. cavernosa* Hoffm.

26) *R. crystallina* L.

27) *R. discolor*Lehm. & Lindenb

28) *R. fluitans* L.

29) *R. frostii* Aust.

30) *R. glauca* Linn.

31) *R. melanospora* Kash.

32) *R. plana* Tayl.

IV] Order: Anthocerotales

xii) Family: Anthocerotaceae *Anthoceros* **L.**

33) *A. bharadwajii* Udar and Asthana

34) *A. crispulus* (Mont.) Douin

35) *A. erectus* Kash.

36) *A. subtilis* Steph.

Phaeoceros **Prosk.**

37) *P. carolinianus* (Michx.) Prosk.

38) *P. himalayensis* (Kash.) Prosk.

39) *P. laevis* (L.) Prosk., *subsp. laevis* Prosk.

Folioceros **Bharad.**

40) *F. dixitianus* (Mahabale) Bharad.,

xiii) Family: Notothyladaceae *Notothylas*

41) *N. indica* Kash.,

42) *N. levieri* St. ex Schiffn.

Amongst all the orders, Marchantialesis the most dominant order in Kolhapur District, followed by Anthocerotales, Metzgeriales and Jungermanniales. Marchantiales with 18 species belonging to 6 genera and 5 families is the largest order followed by Anthocerotales with 10 species belonging to 4 genera and 2 families. The order Jungermanniales and Metzgeriales are represented by 7 species each belonging to 4 genera and 2 families and 5 genera and 4 families respectively. Ricciaceae and Anthocerotaceae are the largest families each with 8 species followed by Lejeuneaceae (5 species), Aytoniaceae (4 species), Aneuraceae and Cyathodiaceae (each with 3 species) whereas Jungermanniaceae, Fossombroniaceae, Targionaceae and Notothyladaceae are represented by 2 species each. Two families *viz*., Metzgeriaceae and Pallaviciniaceae are represented by just a single species (Tables **1** & **2**; Figs. **2** & **3**).

Families with the maximum number of genera were Lejeuneaceae and Anthocerotaceae (Fig. **3**). The most specious and dominant families found in Kolhapur District were Anthocerotaceae, Ricciaceae and Lejeuneaceae followed by Aytoniaceae, Cyathodiaceae and Aneuraceae.

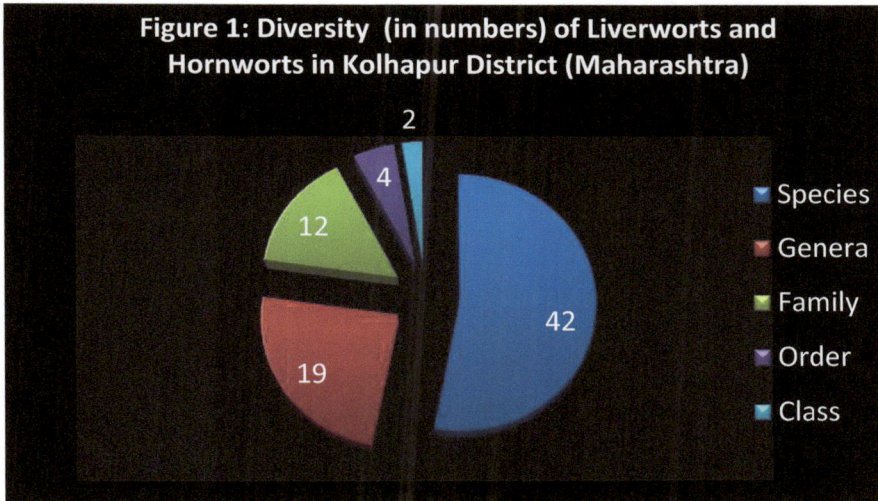

Fig. (1). Diversity of Liverworts and Hornworts from Kolhapur District.

Table 1. Distribution of genera and species of Liverworts and Hornworts of Kolhapur District.

S. No.	Orders	Families	Genera	Species
1.	Jungermanniales	2	4	07
2.	Metzgeriales	4	5	07
3.	Marchantiales	5	6	18
4.	Anthocerotales	2	4	10
Total	4	13	19	42

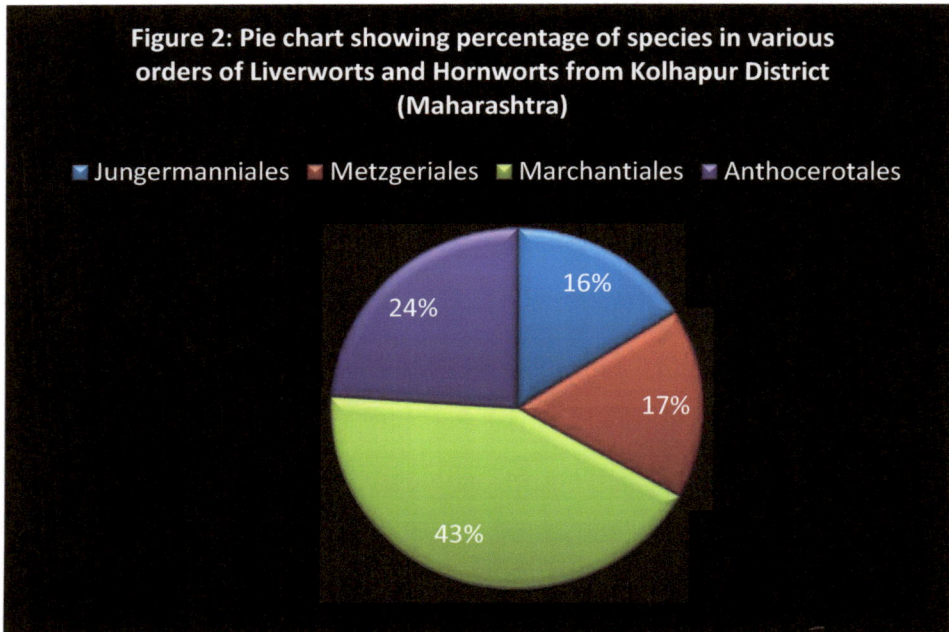

Figure 2: Pie chart showing percentage of species in various orders of Liverworts and Hornworts from Kolhapur District (Maharashtra)

Fig. (2). Pie chart showing number (in %) of species in various orders of Liverworts and Hornworts from Kolhapur District.

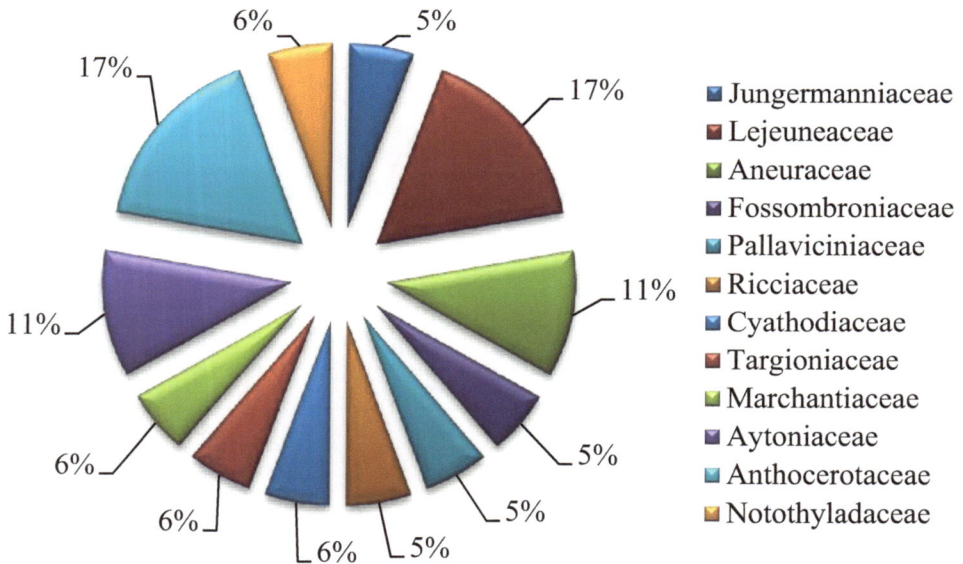

Fig. (3). Species Diversity (in %) in various families of Liverworts and Hornworts.

Table 2. Number of genera and species in each family of Liverworts and Hornworts from Kolhapur District.

S. No.	Order and Family	Number of Genera	Number of Species
I.	**Jungermanniales**		
1.	Jungermanniaceae	1	2
2.	Lejeuneaceae	3	5
II.	**Metzgeriales**		
3.	Aneuraceae	2	3
4.	Metzgeriaceae	1	1
5.	Fossombroniaceae	1	2
6.	Pallaviciniaceae	1	1
III.	**Marchantiales**		
7.	Aytoniaceae	2	4
8.	Cyathodiaceae	1	3
9.	Targioniaceae	1	2
10.	Ricciaceae	1	8
11.	**Anthocerotales**		
12.	Anthocerotaceae	3	8
13.	Notothyladaceae	1	2

Maximum liverworts and hornworts were collected from Chandgad, Panhala, and Gaganbawda where the luxuriant growth of liverworts and hornworts occurs because of high rainfall while minimum from the eastern region of the district like Kagal, some parts of Karveer, Haatkanagle and Shirol (Fig. **4**).

At lower taxonomic level, genus *Riccia* L. with 8 species is the largest genus followed by *Anthoceros* L. (4 species), *Cyathodium* Kunze, *Lejeunea* Lib., *Plagiochasma* Lehm. & Lindenb. and *Phaeoceros* Prosk. (3 species each) and *Fossombronia* Raddi, *Notothylas* Sull., *Riccardia* Gray, *Solenostoma* Mitt., *Targionia* L. (2 species). Seven genera *viz.*, *Aneura* Dumort., *Archilejeunea* (Spruce) Schiffn., *Asterella* P. Beauv., *Cheilolejeunea* (Spruce) Schiffn., *Folioceros* D.C. Bharad., *Marchantia* L., *Metzgeria* Raddi., and *Pallavicinia* Gray, are represented by just a single species.

The species *viz.*, *Anthoceros bharadwajii*, *Cyathodium cavernarum*, *Targionia hypophylla* and *Riccia cavernosa* were collected frequently from the district and were supposed to be abundant in the district while some species like *Solenostoma tetragona, Riccardia santapaui, Fossombronia himalayensis, Metzgeria himalayensis, Marchantia linearis* and *Targionia hypophylla* var. *sinhagarhii* were collected from a single locality and they are presumably rare in the district (Fig **5**).

Bryo-geographical Distribution

Pande [3] on the basis of the distribution and diversity of liverworts and hornworts recognized 7 bryo-geographical regions in India *viz.*, i) the Western Himalaya, ii) the Eastern Himalaya, iii) the Punjab and West Rajasthan, iv) the Gangetic Plains, v) the Central Indian territory, vi) the Western Ghats and vii) the Eastern Ghats and Deccan Plateau. Later Singh [8, 9] proposed the Andaman and Nicobar Islands as a distinct bryogeographical unit which shows a unique bryo-flora considerably different from the mainland. The hepatic flora of the Kolhapur District has elements common to the above territories (Table **3**).

Table 3. Distributional relationship of the Liverworts and hornworts with Indian bryogeographical regions.(As per Pande (1958), Singh (1992, 1997).

Name of the Taxa	Distribution in India							
	*WH	EH	P & WR	GP	CI	WG	EG & DP	A & N
Solenostoma fossombronioides	-	-	-	-	-	+	-	-
Solenostoma tetragonum	-	-	-	-	-	+	-	-
Archilejeunea minutiloba	-	-	-	-	-	+	+	-
Cheilolejeunea intertexta	-	-	-	-	-	+	+	+
Lejeunea discreta	+	+	-	-	-	+	-	-
Lejeunea flava	+	+	-	-	-	+	-	+
Lejeunea tuberculosa	+	+	-	-	-	+	+	-
Aneura pinguis	+	+	+	-	-	+	+	-
Riccardia levieri	+	+	-	-	+	+	-	-
Riccardia santapaui	-	-	-	-	-	+	-	-
Metzgeria himalayensis	+	+	-	-	-	+	-	-
Fossombronia himalayensis	+	+	+	-	+	+	+	-
Fossombronia indica	-	-	-	-	-	+	-	-
Pallavicinia lyellii	+	+	-	-	+	+	+	-
Asterella wallichiana	+	+	+	+	+	+	+	+
Plagiochasma appendiculatum	+	+	+	+	+	+	+	-
Plagiochasma intermedium	+	+	+	+	+	+	-	-

(Table 3) cont.....

Name of the Taxa	Distribution in India							
	*WH	EH	P & WR	GP	CI	WG	EG & DP	A & N
Plagiochasma pterospermum	+	+	+	-	+	+	+	-
Cyathodium cavernarum	+	+	+	+	-	+	+	-
Cyathodium epiphytens	-	-	-	-	-	+	-	-
Cyathodium tuberosum	+	+	+	+	-	+	+	-
Marchantia linearis	-	+	+	-	+	+	-	+
Targionia hypophylla	+	+	+	-	+	+	+	-
Targionia hypophylla var. sinhgarhii	-	-	-	-	-	+	-	-
Riccia cavernosa	+	+	+	+	+	+	+	-
Riccia crystallina	+	+	+	+	+	+	+	-
Riccia discolor	+	+	+	+	+	+	+	-
Riccia fluitans	+	+	+	-	+	+	+	-
Riccia frostii	+	+	+	+	+	+	+	-
Riccia glauca	+	+	+	+	+	+	+	-
Riccia melanospora	+	-	-	-	-	+	+	-
Riccia plana	+	+	+	+	+	+	+	-
Anthoceros bharadwajii	+	+	-	-	-	+	-	-
Anthoceros crispulus	+	+	+	-	-	+	+	-
Anthoceros erectus	+	+	+	-	-	+	+	+
Anthoceros subtilis	-	-	-	-	-	+	-	-
Phaeoceros carolinianus	+	+	+	-	-	+	+	-
Phaeoceros himalayensis	+	+	+	-	-	+	+	-
Phaeoceros laevis subsp. *laevis*	+	+	+	-	-	+	+	-
Folioceros dixitianus	-	-	-	-	-	+	-	-
Notothylas indica	+	-	-	+	+	+	+	-
Notothylas levieri	+	+	-	-	+	+	+	-
Total	31	30	22	12	18	42	27	5

WH: The Western Himalaya; **EH:** The Eastern Himalaya; **P & WR:** The Punjab and West Rajasthan; **GP:** The Gangetic Plains; **CI:** The Central India; **WG:** The Western Ghats; **EG & DP:** The Eastern Ghats and Deccan Plateau; **A & N:** The Andaman and Nicobar Islands.

An analysis of the distribution pattern of the species shows that out of 42 taxa recorded from the study area 7 species are endemic to the Western Ghats of Maharashtra *viz., Solenostoma fosombronioides, S. tetragonum, Riccardia santapaui, Fossombronia indica, Targionia hypophylla* var. *sinhagarhii, Anthoceros subtilis* and *Folioceros dixitianus*. The territory shares the maximum 31 taxa with Western Himalaya followed by 30 with Eastern Himalaya, 27 with Eastern Ghats and Deccan Plateau, 22 with Punjab & West Rajasthan, 18 with Central India each, 12 with Indo-Gangetic plains and 5 with Andaman & Nicobar Islands.

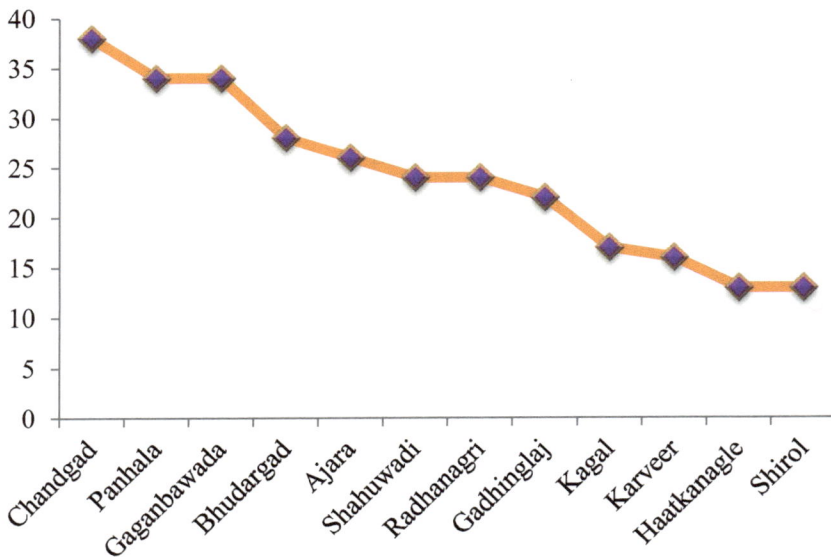

Fig. (4). Tehsilwise number of Liverworts and hornworts species of Kolhapur district (Maharashtra).

Interestingly, none of the species occurs in all the 8 bryo-geographical territories except *Asterella wallichiana*. The maximum bryo-geographical expanse within the country is shown by *Asterella wallichiana* due to its existence in all the territories, followed by *Plagiochasma appendiculatum, Riccia cavernosa, Riccia crystallina, Riccia discolor, Riccia frostii, Riccia glauca* and *Riccia plana* that are distributed in seven studied territories. *Fossombronia himalayensis, Cyathodium cavernarum, Cyathodium tuberosum, Targionia hypophylla, Plagiochasma intermedium, Plagiochasma pterospermum* and *Riccia fluitans* and *Anthoceros erectus* are distributed in six territories, whereas *Aneura pinguis, Pallavicinia lyellii, Anthoceros crispulus, Phaeoceros carolinianus, Phaeoceros himalayensis, Phaeoceros laevis, Notothylas indica* and *Notothylas levieri* are distributed in five territories. Species *viz., Lejeunea flava, Lejeunea tuberculosa* and *Riccardia levieri* are distributed in four territories followed by *Cheilolejeunea intertexta, Lejeunea discreta, Metzeria himalayensis, Riccia melanospora* and *Anthoceros bharadwajii* which are distributed in three territories. *Archilejeunea minutiloba* in two territories while *Solenostoma fossombronioides*, *Solenostoma tetragona*, *Riccardia santapaui*, *Fossombronia indica*, *Cyathodium epiphytens*, *Targionia hypophylla* var. *sinhgarhii*, *Anthoceros subtilis* and *Folioceros dixitianus* were reported only from a single territory.

According to Jog *et al.* [46] and Chaudhary *et al.* [17], based on geographical environment, Kolhapur district can be divided into three zones *viz.*, tropical semi-arid, tropical wet and dry or Savannah and tropical wet.

Tropical Semi-Arid (Steppe) Zone

A long stretch of land situated to the south of Tropic of Cancer and east of the Western Ghats and the Cardamom Hills experiences this climate. It includes Karnataka, Interior and Western Tamil Nadu, Western Andhra Pradesh and Central Maharashtra. This area receives minimal rainfall due to being situated in the rain shadow area. This region is a famine prone zone with very unreliable rainfall varies in between 40 to 75 cm annually (www.wikipedia.org/). This zone includes eastern part of the district, including entire Shirol and Hatkanagle tehsils.

The common liverwort species found in this zone are *Asterella wallichiana*, *Cyathodium tuberosum*, *Plagiochasma appendiculatum*, *P. intermedium*, *P. pterospermum*, *Riccia discolor*, *R. frostii*, *R. melanospora* and *Targionia hypophylla*. Only two hornworts were collected from this zone *Anthoceros bharadwajii* and *A. erectus*.

Tropical Wet and Dry or Savannah Climate

Most of the plateau of peninsula India enjoys this climate, except a semi-arid tract to the east of the Western Ghats. Winter and early summer are long dry periods with temperature above 18°C. Summer is very hot and the temperatures in the interior low level areas can go above 45 °C during May. The rainy season is from June to September and the annual rainfall is between 75 and 150 cm (www.wikipedia.org/). This zone includes Kagal, Gadhinglaj, and Eastern part ofRadhanagari, Bhudargarh, Panhala and Karveer.

The common bryophytes in this zone are *Cheilolejeunea intertexta*, *Cyathodium cavernarum*, *C. tuberosum*, *Lejeunea discreta*, *L. flava*, *Marchantialinearis*, *Plagiochasma appendiculatum*, *P. intermedium*, *Riccia glauca*, *R .cavernosa*, *R. crystallina*, *R. discolor*, *R. fluitans*, *R. frostii* and *R. melanospora*, *Plagiochasma pterospermum* and *Targionia hypophylla*.

However, some common hornworts also collected from this zone are *Anthoceros bharadwajii*, *A. crispulus*, *A. erectus*, and *Notothylas levieri*.

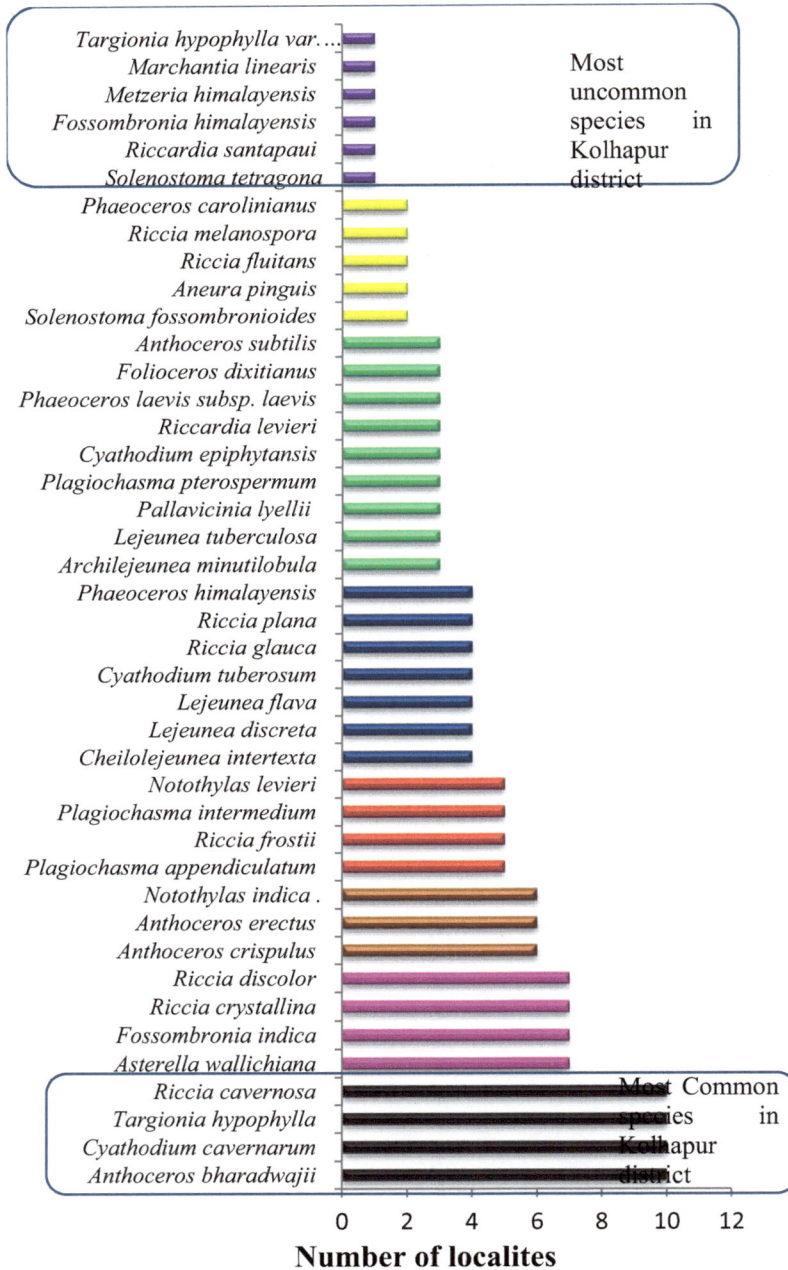

Fig. (5). Species richness of the Liverworts and Hornworts taxa in Kolhapur District.

Tropical Wet Zone

The west coastal lowlands, the Western Ghats, and southern parts of Assam have this type of climate. The rainfall here is seasonal, but heavy. Most of the rain is received in the period from May to November and is adequate for the growth of vegetation during the entire year. However, December - March are the relatively driest months with very little rainfall. The heavy rain is responsible for the tropical wet forests in these regions, which consist of a large number of species of animals. Evergreen forests are the typical feature of the region (www.wikipedia.org/). This zone includes Ajara, Chandgarh, Shahuwadi, Gaganbavada, Western parts of Panhala, Radhanagariand Bhudargarh tehsils.

The common liverworts species found in this zone are *Archilejeunea minutiloba, Aneura pinguis, Asterella wallichiana, Cheilolejeunea intertexta, Cyathodium cavernarum, C. epiphytens, C. tuberosum, Fossombronia himalayensis, F. indica, Lejeunea discreta, L. flava, L. tuberculosa, Marchantia linearis, Metzgeria himalayensis, Pallavicinia lyellii, Plagiochasma appendiculatum, P. intermedium, P. pterospermum, Riccardia levieri, R. santapaui, Riccia cavernosa, R. crystallina, R. discolor, R. fluitans, R. frostii, R. glauca, R. melanospora, R. plana, Solenostoma fossombronioides, S. tetragonum, Targionia hypophylla*, and *T. hypophylla* var. *sinhagarhii.*

The important hornworts include *Anthoceros bharadwajii, A. crispulus, A. erectus, A. subtilis, Phaeoceros carolinianus, P. laevis, P. himalayensis, P. laevis* subsp. *laevis, Folioceros dixitianus, Notothylas indica*, and *N. levieri.*

Distribution Based on Habitat

Habitat wise classification of the Liverworts and hornworts taxa is given in the Table **4**. It seems to be the general rule that liverworts and hornworts flourish well in the crevices where shade and moisture available are more frequent.

Table 4. Distribution of the Liverworts and hornworts in different habitats in Kolhapur District.

Name of the Taxa	Habitat				
	Aquatic	Terrestrial			Epiphytic
		Litho-colous	Terri-colous	Calci-colous	
Solenostoma fossombronioides	-	+	+	-	-
Solenostoma tetragonum	-	+	+	-	-
Archilejeunea minutiloba	-	-	-	-	+
Cheilolejeunea intertexta	-	-	-	-	+

(Table 4) cont.....

Name of the Taxa	Habitat				
	Aquatic	Terrestrial			Epiphytic
		Litho-colous	Terri-colous	Calci-colous	
Lejeunea discreta	-	-	-	-	+
Lejeunea flava	-	-	-	-	+
Lejeunea tuberculosa	-	-	-	-	+
Aneura pinguis	-	+	+	+	-
Riccardia levieri	-	+	+	+	-
Riccardia santapaui	-	+	+	+	-
Metzgeia himalayensis	-	-	-	-	+
Fossombronia himalayensis	-	+	+	-	-
Fossombronia indica	-	+	+	-	-
Pallavicinia lyellii	+	+	+	-	-
Asterella wallichiana	-	+	+	+	-
Plagiochasma appendiculatum	-	+	+	+	-
Plagiochasma intermedium	-	-	-	+	-
Plagiochasma pterospermum	-	+	-	+	-
Cyathodium cavernarum	-	+	+	-	-
Cyathodium epiphytens	-	-	-	-	+
Cyathodium tuberosum	-	+	+	+	-
Marchantia linearis	-	+	+	-	-
Targionia hypophylla	-	+	+	+	-
Targionia hypophylla var. *sinhgarhii*	-	+	+	+	-
Riccia cavernosa	-	+	+	-	-
Riccia crystallina	-	+	-	+	-
Riccia discolor	-	+	+	-	-
Riccia fluitans	+	+	+	-	-
Riccia frostii	-	+	-	-	-
Riccia glauca	-	+	-	-	-
Riccia melanospora	-	+	+	-	-
Riccia plana	-	-	+	-	-
Anthoceros bharadwajii	-	-	+	+	-
Anthoceros crispulus	-	+	-	-	-
Anthoceros erectus	-	-	+	+	-

(Table 4) cont.....

Name of the Taxa	Habitat				
	Aquatic	Terrestrial			Epiphytic
		Litho-colous	Terri-colous	Calci-colous	
Anthoceros subtilis	-	-	+	-	-
Phaeoceros carolinianus	-	-	+	-	-
Phaeoceros himalayensis	-	-	+	-	-
Phaeoceros laevis subsp. *laevis*	-	-	+	-	-
Folioceros dixitianus	-	-	+	-	-
Notothylas indica	-	-	+	-	-
Notothylas levieri	-	-	+	-	-
Total	02	24	27	13	07

Liverworts and hornworts from Kolhapur District can be divided in diverse habitats. The Table **4** shows that 2 species are found in water (aquatic), 32 species are terrestrial out of which, 23 species are found on rocks (Lithocolous), 26 on moist soil floors on clayey slopes on ditches (Terricolous), 13 on brick walls (calcicolous) and 7 species are epiphytic growing on angiosperms. Broadly speaking, thalloid liverworts shows their presence in lithocolous, terricolous and calcicolous habitats except leafy liverworts *Archilejeunea minutiloba, Cheilolejeunea intertexta, Lejeunea discreta, Lejeunea flava, Lejeunea tuberculosa, Cyathodium* spp. and *Riccia fluitans*. Hornworts mostly grow on moist shady places. The increasing urbanization, removal of shady habitats and niches, and changed topography have all posed a threat to its survival.

The liverworts and hornworts grow in an extensive range of habitats. The following types are elaborated on the basis of habitat given by Chaudhary *et al.* [17] and Patil [47].

Open Plateau Region

The zone is fully covered by various types of grasses or herbaceous flora. Only few terrestrial species, *viz. Aneura pinguis, Anthoceros bharadwajii, A. crispulus, A. erectus, A. subtilis, Cyathodium cavernarum, C. tuberosum, Notothylas indica, N. levieri, Riccia billardieri, R. cruciata, R. crystallina, R. discolor, R. fluitans, R. frostii, R. melanospora, R. plana* and *Targionia hypophylla* were found in this region.

Epiphytes

The liverworts and hornworts grow over tree trunks/branches. These are totally autotrophic and only for the sake of shelter they grow on tree trunks. The most common trees in Kolhapur district are *Mangifera indica, Ficus religiosa, Ficus benghalensis, Ficus racemosa, Terminalia* spp., *Syzygium* spp., *Cassia fistula, Bauhinia* spp., and *Dalbergia sissoo* serve as hosts for *Archilejeunea minutiloba, Cheilolejeunea intertexta, Cyathodium epiphytens, Lejeunea discreta, L. flava* and *L. tuberculosa.*

Ravine Liverworts and Hornworts

The liverworts and hornworts along the water channels, streams, rivers, and rivulets among rocks and boulders. The common ravine species are of *Aneura pinguis, Asterella wallichiana, Cyathodium cavernarum, C. tuberosum, Pallavicinia lyellii, Plagiochasma appendiculatum* and *Riccia fluitans.*

Liverworts and Hornworts Near the Domestic Areas

The liverworts and hornworts grow on houses or neighbouring areas. These are *Anthoceros bharadwajii, A. crispulus, A. erectus, C. tuberosum, Notothylas indica, Plagiochasma appendiculatum, P. intermedium, P. pterospermum, Riccardia levieri, Riccia cruciata, R. crystallina, R. discolor, R. frostii* and *Targionia hypophylla* etc.

Liverworts and Hornworts Along the Bridal Path, Roadsides, Border and Cut-edges of the Forests

They are *Anthoceros bharadwajii, A. crispulus, A. erectus, A. subtilis, Asterella wallichiana, Cyathodium cavernarum, C. tuberosum, Fossombronia himalayensis, Folioceros dixitianus, Notothylas indica, N. levieri, Plagiochasma appendiculatum, P. intermedium, P. pterospermum, Riccia billardieri, R. cruciata, R. crystallina, R. discolor, R. frostii, Solenostoma fossombronioides* and *Targionia hypophylla.*

Aquatic Liverworts and Hornworts

Very few species are adapted to aquatic ecosystem e.g. *Pallavicinia lyellii* and *Riccia fluitans.* While, *P. lyelli* is growing on wet rocks at the banks of or into the streams in the rainy season but later it gets exposed. *R. fluitans* is initially floating on the water surface, but gets attached to the substratum.

Endemic Liverworts and Hornworts

Among the species reported from Kolhapur district eight species *viz., Archilejeunea minutilobula, Riccardia santapaui, Fossombronia indica, epiphytansis, Cyathodium epiphytens, Targionia hypophylla* var. *sinhagarhii,*

Riccia plana, Anthoceros bharadwajii, Folioceros dixitianus, Notothylas indica were found to be endemic to the India and Western Ghats.

Habitat destruction creates pressure on the liverworts and hornworts species present in that area. It makes many species threatened, endangered and may finally become extinct. Based on 'The 2000 IUCN World Red List of Liverworts and Hornworts [48] (Craig Hilton-Taylor. 2000) a total of 14 species (34% species) were listed as 'Least Concerned (LC)', whereas 7 species (17% species) were 'Threatened (TH)', 13 species (31% species) were 'Vulnerable (VU)' and 9 species (18% species) were 'not evaluated (NE)' (Fig. **6**).

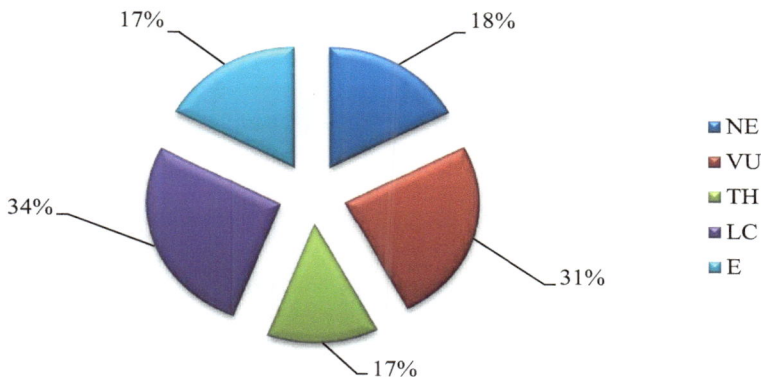

Fig. (6). Pie chart showing the status of species (in %) in Kolhapur district.

In Maharashtra, 14 species of liverworts and hornworts are endemic [26] out of which 6 endemic species *viz.*, *Archilejeunea minutiloba, Riccardia santapaui, Metzgeia himalayensis, Anthoceros bharadwajii, Folioceros dixitianus* are present in Kolhapur district.

Threats to the Wealth of Liverworts and Hornworts from Kolhapur District

Anthropogenic pressure and other biotic activities on the wealth of liverworts and hornworts may lead to decline in their diversity and species may become rare, endangered and threatened (RET).

Forest Fragmentation

Bryophytes including liverworts and hornworts are highly sensitive to the changes in micro-climatic conditions caused by deforestation and environmental pollution. Fragmentation is a dynamic process in which the habitat is gradually abridged into smaller patches that turn out to be more isolated and progressively more affected by edge effects [49 - 53]. Habitat fragmentation is anticipated as one of the main threats to biodiversity. However, the ecological magnitudes of forest fragmentation may depend on the spatial settings of the fragments within the landscape and how these settings change, both temporally and spatially [54]. It is the process of breaking up large patches of forest into smaller pieces. This can be caused by many things, from clearing forest for roads or developmental activities *viz.*, Construction of buildings, colleges, hotels and restaurants. The above activities are common in Amboli, Gaganbavada, Bhudargad, Panhala, Amba, Vishalgad, Radhanagri, Ajara, Pargad and Chandgad. If this process continues, the ability of the remnant forests of Kolhapur District to maintain their original bryodiversity of liverworts and hornworts and ecological processes will be significantly reduced.

Tourisms

Similarly, large number of tourists visiting to Aamba, Radahanagari, Barki Falls, Panhala, Jyotiba, Bhudargad, Vishalgad, Pargad, Shivgad, Paavangad and Kalanidhigad every year. Hence there is tremendous anthropogenic pressure on liverworts and hornworts growing on roadside walls and plateaus.

Cleaning and Rejuvenation of Historical Monuments and Temples

Local peoples, NGO's, government authorities sometimes take steps for cleaning the historical monuments, walls of forts and temples especially on forts causes major threats to the liverworts. Especially as Panhala is a touristic spot the walls of Andharbav, Dhanykothar, Temples, *etc.* were cleaned time to time. Sometimes the crevices of the walls are rejuvenated by filling cement concrete. But the liverworts like *Plagiochasm* sp. and hornwort species like *Anthoceros* sp. growing on mats of mosses on the walls or crevices were always the victims of these practices.

Liverworts *viz.*, *Riccardia levieri*, *Fossombronia indica*, *Riccardia santapaui*, *Riccia fluitans Anthoceros subtilis* and *Folioceros dixitianus* are rare in the study area as they were collected just once during the extensive survey in the Kolhapur District. In addition, there are species like *Riccardia levieri*, *Riccia fluitans* and *Solenostoma fossombronioides* which are confined to just a single location *viz.*, Panhala, Bhudargad and Gaganbavada, respectively within the study area. *Riccia fluitans* was also collected from Panhala in 1998-99, but since 2004 it has been not seen from Panhala but collected from Bhudargad.

Suggestions for the Conservation of Liverworts and Hornworts from Kolhapur District

The species of liverworts and hornworts present inside the Radhanagari Wild Life

Sanctuary automatically gets *in situ* protection. Many forts, Plateaus and peaks of Kolhapur District are also showing some remarkable bryodiversity of liverworts and hornworts. But some unavoidable activities for the better management of these areas, such as construction of inspection paths, roads; cleaning, cementing and coloring of historical monuments and temples on forts and other sites, *etc.*, may cause a threat to the habitats of liverworts and hornworts. To avoid these threats and to ensure more effective conservation of the liverworts and hornworts, the areas in and around the forts *viz.*, Panhalgad, Gaganbavada, Pargad, Shivgad, Bhudargad, Vishalgad and thick forests of Amba, Amboli, Radhanangari and their surrounding areas having luxurious growth may be declared and protected as, "Liverwort and Hornwort Sites", in different altitudinal zones. In Kolhapur District, such bryo-rich areas are Aamba Ghat, Amboli Ghat, Anuskura Ghat, Karul Ghat, Bhudargad, Gaganbavada, Panhalgad, Pargad, Patgaon, Pavangad, Radhanagari WLS, Vishalgad *etc.*, which would also act as 'benchmark' sites for monitoring the health of species populations over a period of time. As the liverworts and hornworts plays an important tool for monitoring the health of forest ecosystems, periodical monitoring of the species populations in these areas would help to assess the impact of various management practices on the overall biodiversity of the Kolhapur District. Besides, these areas can be used for creating awareness among the people, academicians, researchers and the foresters about this interesting group of plants, their role in ecological balance and functions and the threats faced by them. Aamba Ghat, Amboli Ghat, Anuskura Ghat, Karul Ghat, Bhudargad, Gaganbavada, Panhalgad, Pargad, Patgaon, Pavangad, Vishalgad, Panhalgad, Radhanagari WLS are ideal habitats for the growth of liverworts and hornworts. Again Panhalgad and its surrounding area would also be prudent to closely monitor the populations of the species which are represented by just a single or multiple collections from just a single location within the Kolhapur District.

The local inhabitants from the liverwort-hornwort luxuriant areas with interest in environmental conservation may be trained as para-taxonomists to help in the effective monitoring of the species and their habitat on different parts like forts, plateaus and reserve forests in Kolhapur District. Besides, declaring the endemic species as "National Natural Heritage" for their environment in the conservation drive of such taxa of liverworts and hornworts will motivate people residing on the forts, plateaus and reserve forests.

i. There is an urgent need to carry out a systematic bryo-floristic studies of liverworts, hornworts, and mosses from all other districts of Maharashtra.
ii. Identification of bryo-rich areas from Maharashtra having luxuriant growth of liverworts, hornworts, and mosses is necessary to understand their ecology and

their role in the sustenance micro-environment.

iii. It is necessary to ban new constructions of residential, commercial buildings and industries to minimize environmental pollution and interference in the micro-habitats of the liverworts, hornworts, and mosses on the forts, plateaus and forests by strictly enforcing the laws.

iv. Panhalgad and Paavangad are the ideal habitats for the luxuriant growth of the liverworts and hornworts *viz.*, *Aneura pinguis, Anthoceros bharadwajii, A. crispulus, A. erectus, A. subtilis, Asterella wallichiana, Cyathodium cavernarum, C. tuberosum, Folioceros dixitianus, Fossombronia himalayensis, F. indica, Lejeunea flava, L. discreta, Notothylas indica, N. levieri, Phaeoceros carolinianus, P. himalayensis, P. laevis, Plagiochasma appendiculatum, P. intermedium, P. pterospermum, Riccardia levieri, Riccia cavernosa, R. cruciata, R. crystallina, R. discolor, R. fluitans, R. frostii, R. melanospora, R. plana, Targionia hypophylla,* and *T. hypophylla* var. *sinhagarhii.* As these two forts are rich in bryodiversity, it is necessary to declare them as "National Natural Heritage". There is an urgent need to ban new constructions of residential, commercial buildings and industries to minimize environmental pollution and interference in the microhabitats of the liverworts, hornworts, and mosses on the Panhala, Paavangad and other places by strictly enforcing the laws as early as possible.

v. There is an urgent need to conserve the natural habitats of liverworts and hornworts on the forts, plateaus and into the reserve forests of Maharashtra by developing bryophyte gardens under the guidance of expert bryologists of Indian and abroad.

Finally, it is necessary to take initiatives to conserve rare, endangered, threatened, endemic liverworts and hornworts through *ex-situ* and *in-situ* conservative methods in selected areas of Kolhapur, Maharashtra and India.

SUMMARY AND CONCLUSION

The present chapter deals with the study of liverwort and hornwort flora of Kolhapur district with respect to taxonomy, distribution, endemism of liverworts and hornworts species. Comprehensive and extensive collection cum survey tours were conducted from 1997 to 2018 in the widespread localities of Kolhapur district. This has resulted in a collection of 42 species belonging to 19 genera of liverworts and hornworts from Kolhapur district.

For the present study, materials were collected from different vicinities of Kolhapur district. Photographs of liverworts and hornworts were taken in the field itself. These collected species were preserved in 4% formalin. The preserved specimens were deposited in Shivaji University, Herbarium (SUK). The specimens were analyzed, illustrated and identified by using standard literature, protologues, and types.

Taxonomy

Overall, 42 species of liverworts and hornworts have been reported from the area of study. Among these, 33 species belonging to 15 genera and 10 families are of liverworts while 09 species belonging to 04 genera and 02 families are hornworts. Amongst the order Marchantiales had maximum number of genera. It was also the most diversified order, followed by Anthocerotales, Metzgeriales and Junger-manniales. Families with the largest genera were Lejeuneaceae, and Anthocerotaceae. The most diversified families found were Anthocerotaceae, Ricciaceae and Lejeuneaceae followed by Aytoniaceae, Cyathodiaceae and Aneuraceae.

Maximum taxa were collected from Panhala, Chandgad, and Gaganbawda where luxuriant growth of bryophytes occurs because of high rainfall, thick vegetation and high peaks while the minimum from eastern region of the district like Kagal, some part of Karveer, Haatkangle and Shirol due to low rainfall and plains.

Endemic, Rare, Endangered and Threatened Liverworts and Hornworts

Among the studied liverworts and hornworts 17% species are endemic (E), 34% species are least concerned (LC), 17% species threatened (TH), 31% species are vulnerable (VU) and 18% species not evaluated (NE).

The main basic object of investigation is the liverworts and hornworts vegetation and its relationship with environmental factors. It is addressed to the question, *i.e.* how does the species composition react with different environmental factors. During the present investigation 42 were collected amongst these 33 species were found to be terricolous or rupicolous, 7 species were epiphytic and 2 species were aquatic. The most diverse genus is *Riccia* (8 species) and is followed by *Anthoceros* (4 species), *Phaeoceros* (4 species), *Plagiochasma* (3 species), *Cyathodium* (3 species) and *Lejeunea* (3 species). The maximum liverworts and hornworts were reported from tropical wet zone (42 species).

The fast increasing urbanization with its monasteries and recreation on the forts, plateaus and reserve forests is putting extreme pressure on bryo-flora to the limits of their patience. Exploration, collection and conservation are one of the urgent needs of the day.

The extraordinary varied and rich bryo-flora of our vast country is not thoroughly explored till date. Besides Northern-Western Ghats of Maharashtra abounding in moss or liverwort-covered valleys and hillsides, ridges and slopes still await exploration. A great deal of interesting information is shrouded in darkness. Moreover, it has been an urgent call to bryologists and floristic researchers of these areas that un-explored regions deserve first priority for exploration. The need for more and more bryo-exploration in the still very inadequately known parts of the Northern-Western Ghats of Maharashtra is absolutely essential, otherwise many a species would perish and would disappear before being documented form these unexplored areas.

The present work could form a "starting point' and foundation in our region on which more comprehensive qualitative and quantitative studies could rest in the future with the response of liverworts and hornworts to mineral elements and their sources of supply and the factors influencing availability and uptake.

CONSENT FOR PUBLICATION

Not applicable.

CONFLICT OF INTEREST

The authors confirm that this chapter contents have no conflict of interest.

ACKNOWLEDGEMENTS

Author is thankful to Dr. Meena Dongare (Ex Prof. Botany, Shivaji University, Kolhapur),Prin. Rajendra V. Shejwal (LBS College, Satara), Prin. Dr. Anil Patil (SARP Kanya College, Ichalkaranji), Prin. Dr. S.Y. Hongekar (Vivekanand College, Kolhapur), Prin. Milind Hujare (DKASC College, Ichalkaranji) and I/C Prin. Dr. Vitthal S. Dhekale (RR College, Jath) and Dr. S.R Yadav (Ex HOD Botany, Shivaji University, Kolhapur) and Dr. D.K. Gaikwad (HOD Botany, Shivaji University, Kolhapur) for their kind support, constant encouragement and for providing the necessary laboratory facilities to carry out the present research work. Also thankful to Dr. D.K. Singh, Dr. S.K. Singh, Dr. Afroz Alam, Dr. Sachin Patil, Dr. Manoj Lekhak for sparing their valuable time for discussion, generous help, showing interest in criticism and encouragement.

REFERENCES

[1]   Kashyap SR. Liverworts of Western Himalayas and the Punjab Plain, Part I and II (Reprint 1972). Trinagar, Delhi: Researchco Publications 1929-1932.

[2]   Pande SK. Studies in Indian Liverworts. A review. J Indian Bot Soc 1936; 15: 221-33.

[3]   Pande SK. Some aspects of indian hepaticology (Presidential address). J Indian Bot Soc 1958; 37: 221-33.

[4]   Pande SK, Bharadwaj DC. The present position of Indian Hepaticology with a note on the Hepatic vegetation of the country. Palaeobotanist 1952; 1: 368-81.

[5]   Udar R. Bryology in India In: Udar R, Ed. Annales Cryptogamici et Phytopathologici. New Delhi: The Chronica Botanica Co 1976; 4: p. 200.

[6]   Joshi DY, Biradar NV. Studies on the liverwort flora of Western Ghats with special reference to Maharashtra, India. J Hattori Bot Lab 1984; 56: 45-52.

[7]   Asthana AK, Srivastava SC. Indian Hornworts (A Taxonomic Study). Bryophytorum Bibliotheca. Berlin, Stuttgart 1991; 42: pp. 1-230.

[8]   Singh DK. Liverworts (Hepaticae) diversity in India and its conservation.Status Report of Biodiversity Conservation in India. New Delhi: Ministry of Environment and Forests 1992; pp. 1-92.

[9]   Singh DK. Liverworts.Floristic Diversity and Conservation Strategies in India I. 1997; pp. 235-300.

[10]   Singh DK. Diversity in Indian liverworts: their status, vulnerability and conservation. Perspectives in Indian Bryology. Dehradun, India: Bishen Singh Mahendra Pal Singh Publishers 2001; pp. 325-54.

[11]   Singh DK. Notothylaceae of India and Nepal (A morphotaxonomic revision). Dehradun, India: Bishan Singh Mahendra Pal Singh Publisher 2002.

[12]   Joshi DY. A floristic analysis of the liverworts from Andaman Islands, India. Perspectives in Indian Bryology. Dehradun, India: Bishen Singh Mahendra Pal Singh 2001; pp. 135-48.

[13]   Bapna KR, Kachroo P. Hepaticology in India. Delhi: I. Himanshu Publ. 2000.

[14]   Bapna KR, Kachroo P. Hepaticology in India. Delhi: I. Himanshu Publ. 2000.

[15]   Nair MC, Rajesh KP, Madhusoodanan PV. Bryophytes of Wayanad in Western Ghats Malbar Natural History Society. Calicut, Kerala: MNHS 2005.

[16]   Chaudhary BL, Sharma TP, Sandhya C. Bryophyte Flora of Gujarat (India). Udaipur and New Delhi (India): Himanshu Publications 2006.

[17]   Chaudhary BL, Sharma TP, Bhagora FS. Bryophte Flora of North Konkan Maharashtra (India). Udaipur and New Delhi (India): Himanshu Publications 2008.

[18]   Singh AP, Nath V. Heapaticae of Khasi and Jaintia Hills: Eastern Himalayas. Dehradun: Bishan Singh Mahendra Pal Singh Publishers 2007.

[19]   Singh SK, Singh DK. Hepaticae and Anthocerotae of Great Himalayan National Park and Its Environs (HP), India. Kolkata, India: Botanical Survey of India 2009.

[20]   Dey M, Singh DK. Epiphyllous Liverworts of Eastern Himalaya. Thiruvananthapuram, India: Botanical Survey of India 2008.

[21]   Alam A, Srivastava SC. Hepaticae of Nilgiri Hills, Western Ghats (India). Terrestrial Diversity. Germany: LAP-Lambert Academic Publishers 2012.

[22]   Sahu N, Srivastava SC, Alam A. Palyno-taxonomy of some selected taxa of family Aytoniaceae Cavers. Germany: LAP-Lambert Academic Publishers 2013.

[23]   Daniels AED, Daniel P. The Bryoflora of the Southernmost Western Ghats, India. Dehra Dun: Bishen Singh Mahendra Pal Singh Publishers 2013.

[24]   Sandhya Rani S, Sowghandika M, Nagesh KS, Susheela B, Pullaiah T. Bryophytes of Andhra Pradesh. Dehra Dun: Bishen Singh Mahendra Pal Singh Publishers 2014.

[25]   Lavate RA. Studies on the Hepaticae and Anthoerotae of Kolhapur District. 2016.

[26]   Singh DK, Singh SK, Singh D. Liverworts and Hornworts of India: An Annotated Checklist. Kolkata: BSI 2016.

[27]   Manju CN, Rajesh KP. Bryophytes of Kerala. Kozhikode, Kerala, India,: Centre for Research in Indigenous Knowledge, Science and Culture (CRIKSC), 2017; I.

[28]   Lavate RA. Bryodiversity, Distribution, Threats and Conservation of Liverworts and Hornworts from

few forts of Kolhapur District.Biodiversity Assessment: Tool For Conservation. Kolhapur: Bhumi Publishing 2017; pp. 153-66.

[29]   Morajkar AV. Studies on liverworts of Nashik, 1982.

[30]   Kalagaonkar BP. Monographic and Histochemical studies on some hepatic members of Maharashtra. 1990.

[31]   Barve JP. Monographic and Histochemical studies on certain thalloid liverworts from Maharashtra 1990.

[32]   Chavan AR. A new species of *Cyathodium* from India. Bryologist 1937; 40: 57-60. [http://dx.doi.org/10.1639/0007-2745(1937)40[57:ANSOCF]2.0.CO;2]

[33]   Apte VV, Sane PV. Two new species of *Aspiromitus* St. from Bor Ghat. Curr Sci 1942; 11: 59-60.

[34]   Gupte K. Taxonomic observations on a species of *Notothylas* Sull. from Poona. J Univ Bombay 1945; 14: 52.

[35]   Dabhade GT. Mosses of Mahabaleshwar and Khandala with Notes on the Genus Riccia (Mich.) L. in Western Maharashtra 1974.

[36]   Joshi DY. Studies on the bryophytic flora of Western Ghats . 1983.

[37]   Joshi DY. Hepatic flora of the deciduous forest of Purandhar and neighbouring hills, Maharashtra, India. Symp Biol Hung 1987; 35: 515-25.

[38]   Shirke DR. Checklist of bryophytes.Biodiversity of the Western Ghats of Maharashtra: Current Knowledge: 123-130. Dehradhun, India: Bishan Singh and Mahendra Pal Singh Publishers 2002.

[39]   Dabhade GT. Genus Riccia (Mitch.) L. From Maharashtra.Current Trends in Bryology. Dehra Dun, India: Bishen Singh Mahendra Pal Singh Publishers 2007; pp. 255-67.

[40]   Dabhade GT, Hasan A. New species of *Riccia-R. indiragandhi* sp. nov. J Bombay Nat Hist Soc 1986; 83: 398-400.

[41]   Bagawan SA, Kore BA. Liverworts and hornworts of Kas Plateau. Bioscan 2012; 7(2): 289-90.

[42]   Lavate RA. Studies on the liverworts of Panhala 1999.

[43]   Dongare M. An ecological assessment of the liverworts of Panhala hill station (Maharashtra). J Ecophysiol Occup Hlth 2004; 4: 61-6.

[44]   Lavate RA, Patil SB, Shimpale VB, Dongare MM, Patil SM. *Marchantia linearis* Lehm.et Linenb. (Marchantiophyta, Marchantiaceae): A new report from the Western Ghats of Maharashtra, India. DAV Int J Sci 2014; 3(1): 42-6.

[45]   Lavate R, Patil S, Dongare M, Sathe S, Maqdum S. *Pallavicinia lyellii* (Hook.)Gray, (Pallavici-niaceae): An addition to the hepatic flora of Maharashtra, India. Plant Sci Today 2015; 2(4): 192-6. [http://dx.doi.org/10.14719/pst.2015.2.4.167]

[46]   Jog SR, Wakhare A, Chaudhuri S, Unde M, Pardeshi SD. Maharashtra landscape: A perspective.Geography of Maharashtra. Jaipur, New Delhi: Rawat Publications 2002; pp. 19-57.

[47]   Patil SM. Systematic studies on the Pteridophytes of Satara District (Maharashtra) 2014.

[48]   Craig Hilton-Taylor. 2000 IUCN Red List of Threatened Species. The IUCN Species Survival Commission 2000.

[49]   Forman RTT, Godron M. Landscape Ecology. New York: John Wliey 1986.

[50]   Reed RA, Johnson-Bernard J, Baker WL. Fragmentation of a forested rocky mountain landscape, 1950-1993. Biol Conserv 1996; 75: 267-77. [http://dx.doi.org/10.1016/0006-3207(95)00069-0]

[51]   Franklin J. Predicting the distribution of shrub species in southern California from climate and terrain-derived variables. J Veg Sci 1998; 9: 733-48.

[http://dx.doi.org/10.2307/3237291]

[52]   McGarigal K, Cushman SA, Stafford S. Multivariate Statistics for Wildlife and Ecology Research. New York: Springer-Verlag 2000.
[http://dx.doi.org/10.1007/978-1-4612-1288-1]

[53]   Franklin AB, Noon BR, Luke GT. What is habitat fragmentation. Stud Avian Biol 2002; 25: 20-9.

[54]   Drinnan IN. The Search for fragmentation thresholds in a Southern Sydney Suburb. 2005.
[http://dx.doi.org/10.1016/j.biocon.2005.01.040]

CHAPTER 12

Fissidentaceae: A Tiny Fern Moss Family

Mazhar-ul-Islam[*]

Cryptogamic Lab. Department of Botany, Hazara University, Mansehra, Pakistan

Abstract: Family Fissidentaceae is one of the largest family of class Bryopsida. The family is also known as a tinny fern moss family due to the general appearance of plant body similar to that of a typical fern plant. This is a monogeneric taxon (*Fissidens*). This family comprises of ca. 450 taxa distributed all around the world. Among these, 86 reported from India, 53 from china and 37 from North America. Bryoflora of Pakistan represents ca. 18 species of this unique family. The individuals like shady, moist places in forests, entrances of caves, sprays of waterfalls and along rivers, a few species completely aquatic, growing on soil, rocks, termite mounds, lower trunks of trees, branches, dead wood, rarely epiphyllous, in moist or wet lowland to high mountain forests. The identification parameters of this family are somewhat different from others. The identification is based on peristome types, lambidial cells (size, number, and position), laminal cells (size, shape and papillae *etc*.), costa type and in some cases habitats. This chapter focused on the detail background, identification strategies, taxonomic parameters, distribution and conservation strategies *etc*.

Keywords: Fissidentacae, Lambidia, Papilose, Mamilose.

INTRODUCTION

Fissidentaceae is a family of mosses which look like small ferns. The family belongs to Dicranales order with characteristic acrocarpous and haplolepideous mosses [1]. The family is monotypic with a single genus, *Fissidens* Hedw. Morphologically, *Fissidens* are characterized by distichous leaves that are differentiated into a vaginant lamina consisting of two lamellae that clasp the stem, a dorsal lamina (opposite the vaginant lamina), and an apical lamina (above the vaginant lamina). With about 450 species [2], *Fissidens* is one of the largest and most diverse genera of mosses [3], and taxonomically notoriously difficult. Based on the available molecular data, the genus seems to be monophyletic. However, so far, only a few species have been included in phylogenetic reconst-

---

[*] **Corresponding author Mazhar-ul-Islam**: Cryptogamic Lab. Department of Botany, Hazara University, Mansehra, Pakistan; Tel: +92-3343091736; E-mail: mazharawanhu@gmail.com

Afroz Alam (Ed.)

ructions of haplolepideous mosses [4 - 8] and molecular analyses of *Fissidens* are still rare [9, 10].

Distribution

This family comprises of ca. 450 taxa distributed all around the world. Among these, 86 reported from India, 53 from china and 37 from North America. The bryoflora of Pakistan represents ca. 18 species under this unique family. The members of the family are mostly distributed in the humid, warm, tropics of the world [2].

Habitat/ Substrate

The ecological amplitude of the family is very broad; the species grow on a variety of substrates. Members of the family mostly grow in shaded and moist places in forests, entrances of caves; near waterfalls and along the rivers. Some species grow on soil, moist rocks, termite mounds, on tree trunks, branches, dead wood logs, very rare epiphyllous, in moist or wet lowland to high mountain forests and a few species are aquatic (*e.g. Fissidens grandifrons*). They are found from sea level to alpine zones (4450 m altitude from sea level) [2] (Fig. **1**).

Fig. (1). *Fissidens* leaf (a) *Fissidens grandifrons* (b) *Fissidens dubius* (c) *Fissidens ellagens.*

Classification

Fissidentaceae was initially divided into four genera by Brotherus [11, 12] *viz., Simplicidens* Hertz., *Moenkemeyera* C. Muell., *Fissidentella* Cardot., and *Fissidens* Hedw. Currently, Pursell and Bruggeman–Nannenga [13] classified Fissidentaceae to a single genus *Fissidens* and four subgenera *Aloma, Fissidens, Octodiceras* and *Pachyfissidens*, on the basis of peristome type, costa type and number of layers of exothecial cells.

Taxonomy

Fissidens Hedw. is interesting in terms of gametophytic characters, which may be described by the infrageneric classification followed by some workers in the past. Various workers classified species on the basis of peristomial and gametophytic characters as well, while the characters of peristome alone were not enough for the proper taxonomy of the family. A complex taxonomic prototype has led the taxonomists to get multiple approaches towards the identification of species under the genus. Various workers have tried to allot status to the taxa and distribute proper synonyms for them. Significant contributions in this view are those of Bruggeman–Nannenga *et al.* [14], Bruggeman–Nannenga [15], Bruggeman–Nannenga and Roose [16, 17], Bruggeman–Nannenga and Pursell [18] and Stone [19].

Important Taxonomic Characteristics

Gametophyte

Plants in tufts, mats or gregarious, reddish or brownish-green, dull to bright green or blackish green. Stem mostly erect, sometime prostrate, simple or slightly branched (Fig. **2**).

Fig. (2). A fertile plant of *Fissidens* sp.

Leaf

Leaves distichous are only found in Fissidentaceae. In this case, the leaf

consisting of 3 parts *viz.*, conduplicate or vaginent lamina, apical lamina and dorsal lamina [2] with single costa, usually strong, sometimes branched at apex; apical and dorsal lamina with unistratose or pluristratose margin; Gemmae mostly absent rarely present on the leaf tips, axils or on rhizoids (Fig. **3**).

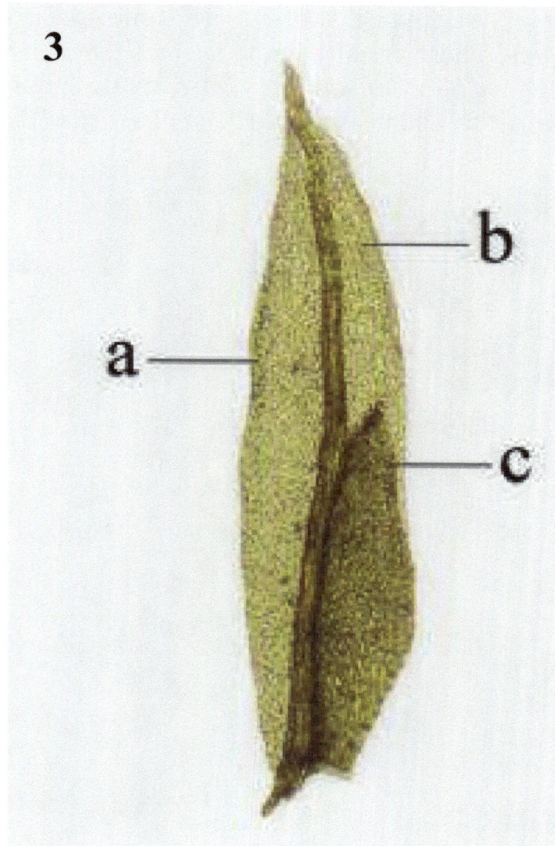

Fig. (3). *Fissidens* leaf ; distichous type of leaf (**a**) Dorsal lamina, (**b**) Apical lamina, (**c**) Vaginant lamina [10].

The vaginant lamina and apical lamellae clasp the stem [20]. This character can only be seen in *Fissidens* which becomes one of the key characters in separating this genus from other genera. The vaginant laminae may be equal or unequal in size, but most of the species having unequal vaginant laminae [2]. This unique character is also helpful in differentiation of some of the species this family.

The leaves of *Fissidense* are distichous, arranged in two opposite rows, vertically oriented on the stem in the equatorial plane and clasping the stem [2]. The pattern of leaves can be used to distinguish *F. flaccidus* and *F. bogoriensis* from other species as their leaves are drooping as compared to other leaves which are generally rigid. Leaf shape for *Fissidens* species sharp. Most of them are lanceolate to narrowly lanceolate. Lanceolate leaves can be seen in *F. hollianus* and *F. pellucidus*. There are also elongate lanceolate leaves which can be seen in *F. nobilis* some others. Other than that, there are some species with lanceolate to oblong lanceolate leaves. There are also some species that can be distinguished by their ovate leaves such as *F. subdiscolor*. While, ovate lanceolate to lanceolate leaves can be found on a few species such as *F. ceylonensis* (Fig. **4**).

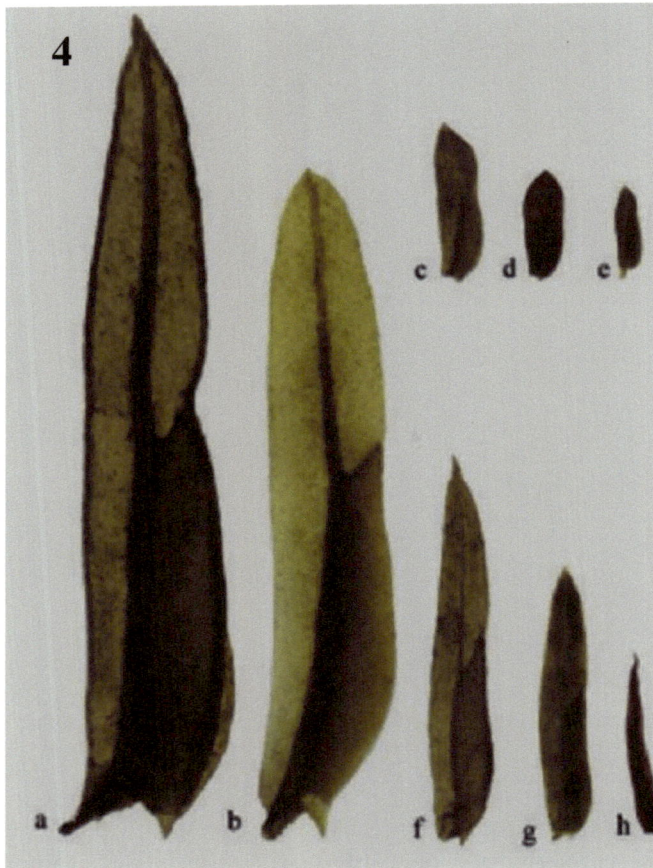

Fig. (4). Leaf shape (**a**) *Fissidens nobilis* Griff. (**b**) *F. polypodioides* Hedw. (**c**) *F. hollianus* Dozy and Molk. (**d**) F. punctulatus Sande Lac. (**e**) *F. ceylonensis* Dozy and Molk. (**f**) *F. javanicus* Dozy and Molk. (**g**) *F. crispulus* var. crispulus and (**h**) *F. crassinervis* Sande Lac. [10].

Apex

The apex of the leaves varies from species to species, ranging from narrowly or widely acute to obtuse leaf apices. This character is useful to distinguish some species such as *F. subdiscolor* which have round to obtusely acute apices. *Fissidens crispulus* var. *crispulus* also can be distinguished by its broadly acute-mucronate leaf apices [21]. Some species also can be differentiated by the shape of apices in dry condition, such as *F. leptopelma* and *F. crispulus* var. *robinsonii*. Both species have narrow acute leaf apices, but *F. leptopelma* differs in the strongly curled apices when dry [21]. The narrowly acute leaf apices are usually used to differ *F. crassinervis* from other species. The leaf apex for *F. nobilis* is acute-subacute with mucronate tips and the costa is shortly excurrent. Round to widely obtuse apices are used to identify *F. punctulatus* [21] (Fig. **5**).

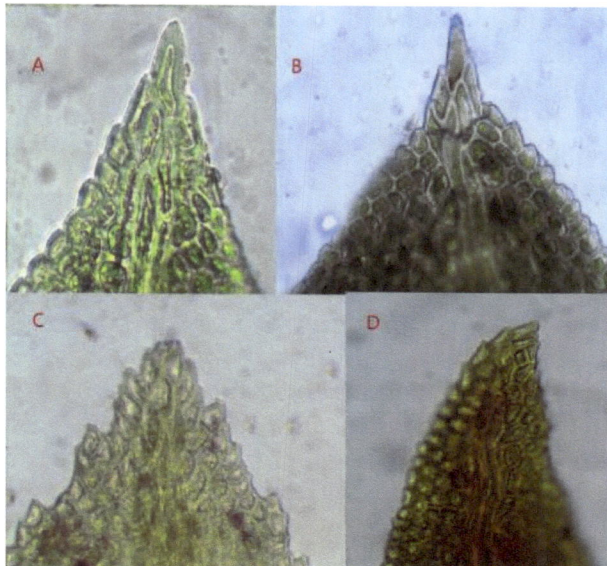

Fig. (5). Variation in leaf apices of Fissidentaceae.

Leaf Base

The decurrent leaf base is found in Fissidens taxa such as in *F. geminiflorus* var. *geminiflorus*. This clearly wide decurrent leaf base in an undulate wing in this species helps in differentiating *F. geminiflorus* var. *geminiflorus* from other species [22]. Also, the ending of dorsal lamina of *F. crassinervis* is usually reduced to a single row well above the base of costa can help to distinguish this species from *F. pellucidus*, wherever the dorsal lamina usually ending more or less abruptly at the base of costa [22].

Leaf Margins

Leaves margins varying from crenulate, entire, denticulate or serrulate to irregularly serrate or limbate with single costa, usually strong, sometimes branched at apex; most of them are entire-serrate. For instance, *Fissidens taxifolius* is distinguished by its evenly serrulate or crenulate-serrulate leaf margin, while F. elegance serrulate and elimbate, in case of *F. dubius* serrpate to crenulate-serrulate but irregularly serrate at leaf apex, elimbate; while in case of *F.grandifronds*, margins are slightly dentate or serrulate; *Fissidens massureinsis,* margin dentate at apical and dorsal lamina but serrate at vaginant lamina; in case of *F. viridulus,* margins slightly denticulate or entire, the margin of *F. bryoides* is entire and limbate on all laminae (dorsal, apical & vaginant).

F. flaccidus showed a more or less entire leaf margin as well as *F. zollingeri*. *Fissidens flaccidus* also showed a somewhat serrulate leaf margin. Also, *F. guangdongensis* has serrulate to almost entire leaf margins which can be used to diverge it from other species. The difference of leaf margin between *F. nobilis* and *F. javanicus* help to differ them as the leaf margin for the latter species is crenulate towards apex whereas the leaves of *F. nobilis* is clearly toothed towards apex. The margin also sometimes varies at the apical, dorsal and vaginant laminae. Generally, margin for vaginant lamina is not the same to the other lamina. This character can help to distinguish the species, for example, the leaf margin of vaginant lamina in *F. serratus* is strongly serrate which helps to demarcate the species. *Fissidens papillosus* also have distinctly spinose-serrate margin of vaginant lamina which also help in delimiting the species. The marginal cell around the leaf is not differentiated in elimbate leaves, while there are some species that have leaves with a thick, dark margin composed of two and more layers of short cell, such as *F. javanicus*. This is similar to the margin in *F. nobilis*. In contrast, there are also some species without the dark and thick margin. Species such as *F. crispulus* var. *robinsonii* have perichaetial leaves where the lower marginal cells of vaginant lamina are larger and pellucid; elongate to rectangular which form an indistinct border. This character has been used to differentiate this species [22].

The leaf margin, which consists of a band of uni-stratose to pluri-stratose, hyaline-yellowish, prosenchymatous, stereid cells is called limbidium [2]. In some species, the limbidium can be well developed in all three laminae. There are some species which show this character such as *F. benitotanii, F. bryoides* var. *ramosissimus. F. flaccidus* and *F. zollingeri.* However, most of the species *Fissidens*, the limbidia are limited to a certain part of the leaves, which is generally the vaginant laminae and often to perichaetial leaves such as in *F. ceylonensis, F. hollianus* and *F. wichurae.* The limbidia also can be intra-laminar in which it is delimited by one or more rows of chlorophyllose laminal cell, such as *F. serratus* [2] (Fig. **6**)

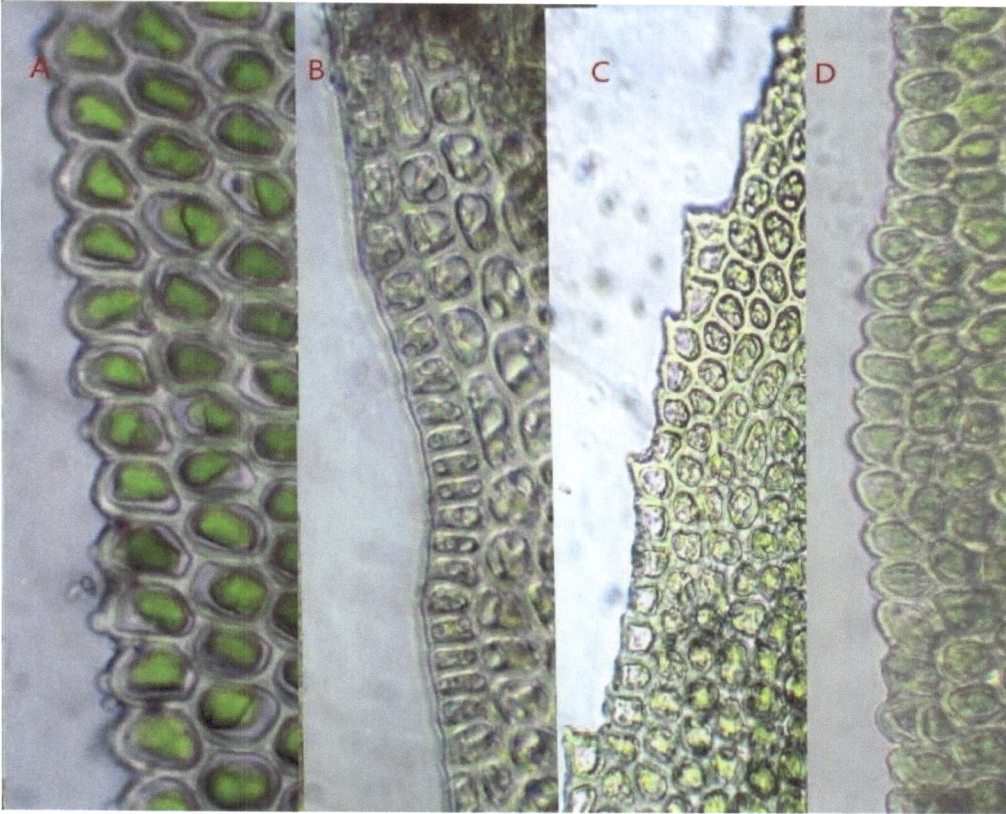

Fig. (6). Variation in leaf margins of Fissidentaceae.

Leaf Cells

The laminal cells are more or less hexagonal to excurrent in short apiculus; Median and upper Laminal cells are oblong lanceolate and elliptic to lanceolate, really ligulate.

Laminal cells, of *Fissidens* species are variable, thin or thick-walled, smooth, papillose or mamillose [22]. In various species such as *F. bogoriensis* and *F. flaccidus*, their apical laminal cells are lax and large, elongate and thin-walled. The thick cell wall of apical lamina in *F. pellucidus* and *F. crassinervis* helps in distinguishing both the species from the others. Smooth laminal cell can be seen in species such as *F. crassinervis, F. flaccidus*, and *F. zollingeri*. Definitely uni-papillose laminal cells of *F. punctulatus* and *F. papillosus* can be used to separate both species from the others. There are also some species that can be known by their pluri-papillose laminal cell such as *F. ceylonensis* and *F. hollianus*.

Fissidens crenulatus var. *elmeri* can be different from other species by the mamillose laminal cells with one high papilla [23] (Fig. **7**).

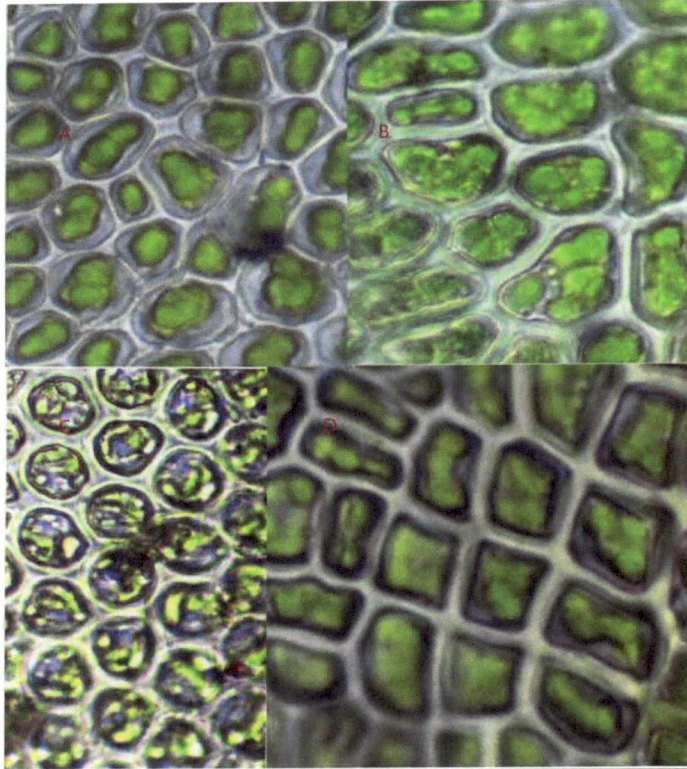

Fig. (7). Variation in leaf cells of Fissidentaceae.

Papillae

Many theories have been proposed for papillae, but little is available as experimental evidence to support them. The papillae have a number of shapes and forms, while varying in size and density. Based on this variability, it seems that their functions will be variable in all species or under all conditions.

Costa

Costa is ending a few cells before the apex can be seen on leaves of *F. guangdongensis* and *F. flaccidus*. A few species have percurrent-shortly excurrent costa. Besides, the absence of costa in the leaves will differentiate species.

Axillary Nodules

Hyaline nodules are to be found in the axils of the leaves. This character has been considered and discussed by Iwatsuki and Pursell [24]. Morphologically axillary hyaline nodule is a branch primordium [2]. These structures are also called by other terms such as 'clusters of hyaline cells' [21], "axillary glandular structures" [25] and "clusters of enlarged cells" [26]. This character can be used to differentiate between *F. bryoides* var. *ramosissimus* and *F. zollingeri* as both taxa are similar in terms of having limbidia on apical, dorsal and vaginant laminae. This character can also be used to differentiate *F. crispulus* var. *crispulus* and *F. crispulus* var. *robinsonii* from other species. Hence, this character helps to differentiate this species from *F. oblongifolius* as it does not possess axillary hyaline nodules. *F. javanicus* can be separated from *F. nobilis* by the presence of clearly differentiated axillary hyaline nodules.

The size of the axillary cell also helps in delimiting the taxa, *F. geminiflorus* var. *geminiflorus* where the axillary nodules consist of smaller cell than the axillary nodule of *F. crispulus* var. *crispulus* and *F. javanicus*. (Fig. **8**).

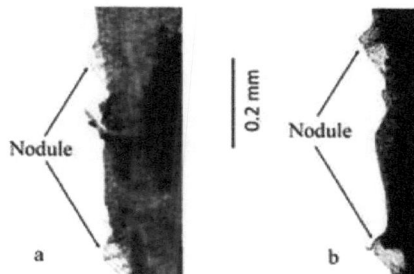

Fig. (8). Axillary hyaline nodules **(a)** *Fissidens crispulus* var. crispulus and **(b)** *F. javanicus* Dozy and Molk [10].

Axillary Hairs

These are derived from epidermal cells; these are uni-seriate and while they are ephemeral, are clearly observed in the axils of distal leaves [2]. Axillary hairs are used by Saito [3] at the generic level. Saito has documented two types of axillary hairs based on differences in the basal cell of the hair. In the first type, the basal cell is shorter and pigmented while the lasting cell are larger and hyaline. In the second type, all cells are hyaline and more or less alike. The axillary hairs of *Fissidens* fall in the second type; however, their lengths are different species to species [2].

Sporophyte

Sporophyte terminal or lateral; Capsule inclined or straight, symmetric or asymmetric, ovoid to cylindrical in shape; Operculum conical, short or long rostrate; Peristome single, reddish to brownish with 16 teeth; Calyptra mitrate or cucullate, rough or smooth; Spores oblate, smooth or papillose.

Seta

Seta in *Fissidens* is usually 2 to some mm long, smooth or papillose [22]. This character is useful in distinguishing the species as the softly scabrous seta in *F. hollianus* is used to differentiate this species from *F. ceylonensis* and *F. wichurae* which have smooth setae [23]. *Fissidens crassinervis* is also characterized by having smooth seta, which helps to distinguish this species from the others (Fig. **9**).

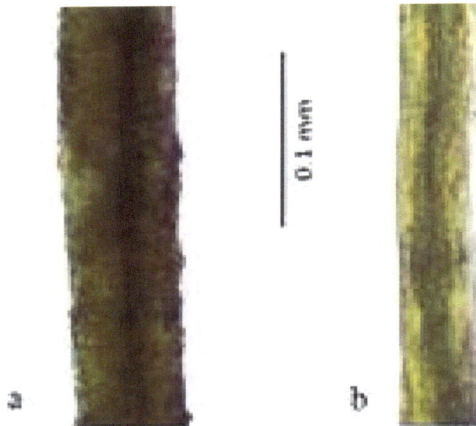

Fig. (9). Seta **(a)** *Fissidens hollianus* Dozy and Molk. and **(b)** *F. ceylonensis* Dozy and Molk [10].

Peristome

The capsule of *Fissidens* has an endostome of 16 haplolepideous teeth, each of them generally divided 1/2-2/3 its length into two filaments of more or less equal lengths [2]. In an SEM study of variations in the peristome of *Fissidens*, Allen [27] has distinguished seven types among 19 species on the basis of differences in the trabeculae and lamellae of the dorsal and ventral surfaces. Then, Stech *et al.* [28] study of peristome types of *Fissidens* in which they have recognized that there are five basic peristome types, *i.e.* bryoides, scariosus, similiretis, taxifolius and the zippelianus type.

Sexuality

Fissidens species are both monoicous and dioicous. There are a few variations under the monoicous such as cladautoicous, gonioautoicous, polyoicous rhizautoicous and synoicous while there is only one difference in dioicous which is pseudautoicous. This character can help in distinguishing species such as *F. pellucidus* and *F. crassinervis* where the lateral is dioicous and the former is synoicous [2].

Beside of above characteristics, fissidentaceae is also characterized by calyptras types. However, molecular investigations are needed for the better studies of this family.

CONCLUSION

Fissidentaceae is one of the most intricate family of mosses and distributed well in Pakistan and elsewhere. The morphological characteristics are quite unique, especially the foliage. Peristome is another key factor which is helpful in the identification of mosses under this family. This outcome presented here is largely the product of the author's own research with some inputs of other researchers working on this family. Images used in this chapter were snapped during the PhD work of principle author (MA) and Dr. Nik Norhazrina, Malaysia. An attempt has been made to discuss all important taxonomic features of Fissidentace along with relevant images. This chapter will be helpful for the readers who are engaged in morpho-taxonomy of the bryophytes.

GLOSSARY

Acrocarpous	The moss plants bearing the archegonia and antheridia at tip of the main stem.
Acute	Tapping with more or less straight margins.
Apical lamina	The part of leaf in Fissidentaceae which is at the apex of leaf.
Apiculus	A short abrupt point at the tip or apex of leaf or sporophyte.
Crenulate	Having minute rounded teeth or scal-lops along the edge usually the bulging walls of individual cells.
Denticulate	Leaves with fine teeth often just cell tips projecting from the margin.
Distichous leaf	Leaves arranged in two rows or ranks on opposite sides of a stem.
Dorsal lamina	In *Fissidens* leaf the wing opposite the sheathing base or vaginant lamina.
Entire	The margin which is smooth, lacking any teeth, cilia, indentations or fingers etc.
Gregarious	Plants growing closely but not very dense.
Haplolepideous	In mosses a type of peristome having only one ring of teeth.
Lamella	A wall-like rib or flap running lengthwise down the leaves of some mosses and the thalli.

Limbidium	In Mosses leaves long border cells.
Lingulate	Tongue shaped or strap-shaped.
Mamilose	Bulging with a blunt central projection.
Mat	A growth of moss in which stems are flattened on the substratum and densely interwoven.
Monotypic	A taxonomic rank that contains only one member in the rank below it, *e.g.* a family with one genus or a genus with one species.
Papillae	Singular papilla; a solid minute pro-tuberance on the cell surface.
Serrate	Regular toothed like a saw blade with teeth composed of one or more cells and pointing towards the apex.
Serrulate	Minutely regularly toothed with teeth composed of only part of a single cell.
Tufts	A clump of more or less erect shoots.

CONSENT FOR PUBLICATION

Not applicable.

CONFLICT OF INTEREST

The authors confirm that this chapter contents have no conflict of interest.

ACKNOWLEDGMENTS

The author is thankful to Dr. Nik Norhazrina, Senior Lecturer, Faculty of Science and Technology, Universiti of Kebangsaan, Malaysia for kind permission to use some images. Thanks are also gratitude to Prof. Dr. Manzoor Hussain for reviewing the manuscript before submission. The author is also thankful to Dr. Jan Alam, Assistant Professor, Department of Botany, Hazara University, Mansehra for his critical attention to this manuscript.

REFERENCES

[1]   Frey W, Stech M. Marchantiophyta, Bryophyta, Anthocerotophyta 2009.

[2]   Pursell RA. Fissidentaceae − Fl. Neotrop Monogr 2007; 101: 1-278.

[3]   Saito K. A monograph of Japanese Pottiaceae (Musci). J Hattori Bot Lab 1975; 39: 373-537.

[4]   Cox CJ, Goffinet B, Wickett NJ. Moss diversity: A molecular phylogenetic analysis of genera. Phytotaxa 2010; 9: 175-95.
      [http://dx.doi.org/10.11646/phytotaxa.9.1.10]

[5]   Fedosov VE, Fedorova AV, Fedosov AE, Ignatov MS. Phylogenetic inference and peristome evolution in haplolepideous mosses, focusing on Pseudoditrichaceae and Ditrichaceae s.l. Bot J Linn Soc 2016; 181: 139-55.
      [http://dx.doi.org/10.1111/boj.12408]

[6]   Hedderson TA, Murray DJ, Cox CJ. Phylogenetic relationships of haplolepideous mosses (Dicranidae) inferred from rps4 gene sequences. Syst Bot 2004; 29: 29-41.
      [http://dx.doi.org/10.1600/036364404772973960]

[7]   La Farge C, Misher BD, Wheeler JA. Phylogenetic relationships within the haplolepideous mosses. Bryologist 2000; 103: 257-76.
[http://dx.doi.org/10.1639/0007-2745(2000)103[0257:PRWTHM]2.0.CO;2]

[8]   Stech M, McDaniel SF, Hernández-Maqueda R. Phylogeny of haplolepideous mosses – challenges and perspectives. J Bryol 2012; 34: 173-86.
[http://dx.doi.org/10.1179/1743282012Y.0000000014]

[9]   Werner O, Patiño J, González-Mancebo JM. The taxonomic status and the geographical relationships of the Macaronesian endemic moss Fissidens luisieri (Fissidentaceae) based on DNA sequence data. Bryologist 2009; 112: 315-24.
[http://dx.doi.org/10.1639/0007-2745-112.2.315]

[10]   Syazwana N, Norhazrina N, Maideen H, Yong KT, Suleiman M. Towards a revision of the moss genus (Fissidentaceae) in Peninsular Malaysia. Malay Nat J 2018; 70(3): 297-307.

[11]   Brotherus VF. Fissidentaceae.edited by Engler A and Prantl K 1: 351-363. Leipzig, Germany 1901.

[12]   Brotherus VF. Musci.edited by Engler A and Prantl K 10: 243-302. Leipzig, Germany 1924.

[13]   Pursell RA, Bruggeman-Nannenga MA. A revision of the infrageneric taxa of Fissidens. Bryologist 2004; 107: 1-20.
[http://dx.doi.org/10.1639/0007-2745(2004)107[1:AROTIT]2.0.CO;2]

[14]   Bruggeman-Nannenga MA, Pursell RA, Iwatsuki Z. A re-evaluation of Fissidens subgenus Serridium section Ambylothallia. J Hattori Bot Lab 1994; 77: 255-71.

[15]   Bruggeman-Nannenga MA. Notes on Fissidens VI. New synonyms, new combinations & validation of some names. J Hattori Bot Lab 1997; 81: 155-73.

[16]   Bruggeman-Nannenga MA, Roose MC. On the peristome types found in the Fissidentaceae and their importance for the classification. J Hattori Bot Lab 1990; 68: 193-234.

[17]   Bruggeman-Nannenga MA, Roose MC. Cladistic Relationships between the main peristome types of the Fissidentaceae. J Hattori Bot Lab 1990; 68: 235-8.

[18]   Bruggeman-Nannenga MA, Pursell RA. Notes on Fissidens V. Lindbergia 1995; 20: 49-55.

[19]   Stone IG. Fissidens linearis Brid. and its synonyms. J Bryol 1991; 18: 159-67.
[http://dx.doi.org/10.1179/jbr.1994.18.1.159]

[20]   Bruggeman-Nannenga MA. Subgenus Fissidens in tropical Eastern Africa with emphasis on the Tanzanian collections by Tamás Pócs. Pol Bot J 2013; 58(2): 369-417.
[http://dx.doi.org/10.2478/pbj-2013-0055]

[21]   Iwatsuki Z, Suzuki T. Fissidens in the Ryuku Islands, Japan. J Hattori Bot Lab 1977; 43: 379-408.

[22]   Eddy A. A Handbook of Malesian Mosses 1 Sphagnaceae to Dicranaceae. London: Natural History Museum Publications 1988.

[23]   Iwatsuki Z, Mohamed MAH. The genus Fissidens in Peninsular Malaysia and Singapore (a preliminary study). J Hattori Bot Lab 1987; 62: 339-60.

[24]   Iwatsuki Z, Pursell R. Axillary hyaline nodules in Fissidens (Fissidentaceae). J Hattori Bot Lab 1980; 48: 329-35.

[25]   Norkett AH. Some problems in a monographic revision of the genus Fissidens with special reference to the Indian species. Bulletin of the Botanical Society of Bengal 1969; 23: 75-82.

[26]   Robinson H. Observations on the origin of the specialized leaves of Fissidens and Schistostega. Rev Bryol Lichenol 1970; 37: 941-7.

[27]   Allen BH. Peristome variations in the genus Fissidens: An SEM study. Bryologist 1980; 83: 314-27.
        [http://dx.doi.org/10.2307/3242441]

[28]   Stech M, McDaniel SF, Hernández-Maqueda R. Phylogeny of haplolepideous mosses – challenges and perspectives. J Bryol 2012; 34: 173-86.
        [http://dx.doi.org/10.1179/1743282012Y.0000000014]

SUBJECT INDEX

www.ingramcontent.com/pod-product-compliance
Lightning Source LLC
Chambersburg PA
CBHW041701210326
41598CB00007B/488